OF SHIPS AND ISLANDS

OF SHIPS AND ISLANDS

Stuart S Walker

Stuart S. Walker

The Book Guild Ltd
Sussex, England

First published in Great Britain in 2004 by
The Book Guild Ltd
25 High Street
Lewes, East Sussex
BN7 2LU

Typesetting in Times by
SetSystems Ltd, Saffron Walden, Essex

Printed in Great Britain by
Antony Rowe Ltd, Chippenham, Wiltshire

A catalogue record for this book is
available from the British Library

ISBN 1 85776 815 9

CONTENTS

MAPS

AUTHOR'S NOTE

While the names of some of the places in this book have changed in recent years, I have referred to them by the names that were in use at the time of my working there.

1

Maiden Voyage

It was still dark on a cold January morning as I waved
goodbye to my mother and crossed the road at Cardwell
Bay to catch the 7.00a.m. Gourock to Glasgow bus – the
bus that would take me on the first part of the long journey,
by way of the Bay of Biscay and Suez, then on past Aden,
to Rangoon. The bus arrived dead on time and soon I was
sitting peering nostalgically through the window as the
familiar scenes of Greenock and Port Glasgow – Kincaid's
marine engine works, the James Watt Dock, Lithgow's
Shipyard and the Gourock Ropework – slid past, dimly
illuminated by the street lighting. Dawn was breaking as we
passed Langbank with its long mud-flats uncovered by the
low tide; then it was up the hill towards Erskine and
Renfrew, with the daylight growing stronger every minute.
I looked back through the rear windows to catch a last
glimpse of one of my favourite views: looking down the
river, past Dumbarton and Langbank to the broad reach of
the Clyde at the Tail of the Bank with the Argyll hills
beyond. That morning the early winter sun was just beginning to give a pink tinge to the snow-covered hills.

Soon we were in Glasgow and I was alighting at a bus
stop in the Paisley Road. From there it was just a short
walk down a side street to reach Plantation Quay where my
ship, Paddy Henderson's *Bhamo*, was lying loaded and
ready to sail for Rangoon. Everything in the engine-room

had been checked and made ready the day before, sea watches had started at 8.00a.m. and we were all set to go. There was nothing for me to do that morning, except, of course, to double-check everything once again – if something is going to go wrong, you can bet your bottom dollar it will do so at the most inconvenient moment.

Having previously served on several of Henderson's 'K' class cargo ships on the West African service, I had been appointed to *Bhamo* as 2nd engineer in November 1957, while she was in the final stages of fitting-out in the James Watt dock at Greenock. *Bhamo* was the latest addition to the fleet of P. Henderson & Company Limited, 95 Bothwell Street, Glasgow, familiarly known to the maritime world as Paddy Henderson. Initially formed in 1835, the company had in the early days employed chartered ships on the short sea-routes to Europe. Later they operated their own ships trading as far afield as Australia and India. But it was with New Zealand and Burma that the Henderson Line ships were to be mainly involved. From 1854 they began trading their sailing ships down to New Zealand as the Albion Line. The trade to New Zealand in those early days was mainly one-way, with settlers and manufactured goods from Glasgow. However, Henderson's ships found a profitable return business by running in ballast up to Moulmein and Rangoon to load rice and teak for British and continental ports. The success of this trade led Henderson to open a direct regular service between Glasgow and Burmese ports in 1860. From 1873 three companies were engaged in the New Zealand trade – Henderson's Albion Line, Shaw Savill of London and the recently formed New Zealand Shipping Company. Competition was stiff and in 1882, in order to provide a better service and counter the threat from the rival New Zealand Shipping Company, the two older companies amalgamated to form the famous Shaw Savill & Albion Line. Thus ended Henderson's direct involvement in the New

2

Zealand trade; but not before their sailing ship *Dunedin* made history by being the first ship to bring a cargo of frozen New Zealand lamb to the British market. She sailed from Port Chalmers on 15 February 1882 and arrived at London with the cargo in excellent condition on 26 May. A whole new industry had been created.

P. Henderson & Co. then concentrated on the service to Burma and, gradually, all the ships came to be named after the towns and rivers of Burma. With the decline of the Burma trade in the post-independence world, most of the Henderson cargo ships were chartered to Elder Dempster, who found them very suitable for their West African service. Only *Prome*, *Salween*, *Martaban* and *Yoma* remained on the service to Rangoon, to be joined in December 1957 by the newly delivered *Bhamo*, named after an important trading centre on the upper Irrawaddy, near the border with China.

The 'K' class ships that I had previously served on had been engaged on Elder Dempster's West African service and had Nigerian crews; but *Bhamo* was designed for Paddy Henderson's traditional Burma service and had a Burmese crew. They were absolutely first class. They joined the ship at Plantation Quay just after we took delivery from the shipbuilders, Lithgow's of Port Glasgow, on 28 December. The state of the engine-room at that time was rather chaotic, with spare parts, tools and general debris left behind by the ship-builders, scattered about here and there. They arrived about 10.00a.m. and I indicated to the Serang, the senior engine-room rating, that, after they had stowed their gear and sorted themselves out, they should start to tidy the place up. I was kept busy outside the engine-room for most of the day and it was about 3.00p.m. when I went back to see how things were progressing down below. The engine-room was immaculate, not a thing out of place. I had never seen such a transformation in such a short time.

3

Just before noon, with tugs assisting fore and aft, *Bhamo* moved away from the quay out into mid-stream and started moving slowly down river. It was the start of her maiden voyage from Glasgow to Rangoon. On the bridge the pilot nodded and said, 'Slow-ahead will do her now, Captain'. Our 'Old Man', Piggy Walker, swung the telegraph handles to 'slow-ahead'. We were on our way.

Prior to that moment we had been on 'stand-by' conditions down in the engine-room for the best part of an hour with Colin Kerr, the chief engineer, pacing nervously up and down the floor plates. The 3rd engineer had been hovering over his generators, while the 4th was making sure his compressors were giving of their best, and the electrician was gazing, as if hypnotised, at the main switchboard instrumentation and fiddling with his circuit-breakers and voltage regulators. I had been standing at the manoeuvring platform, scanning the various main-engine temperature and pressure gauges for the umpteenth time, making sure that everything was ready for the start of the voyage and answering Colin Kerr's numerous, somewhat agitated, enquiries as to our state of readiness. Suddenly, the engine-room telegraph on my left-hand side clanged into life. The pointer swung back and forward round the dial and stopped at 'slow-ahead'. I acknowledged the order, moved the engine control lever to 'start' and then to the 'slow-ahead' position. The engine hissed into life, then settled down to its rhythmic, slow-speed, diesel beat of about 50 revs per minute as we nosed our way down the Clyde.

We had three or four technicians from the engine builders, John G. Kincaid of Greenock, on board to monitor and record the main-engine performance with the ship in a loaded condition. To this end, after we had completed the down-river passage, we carried out half a dozen full-speed runs on the measured mile off Skelmorlie. Then, after they had collected all the necessary data, Kincaid's men disembarked, together with the pilot, at Gourock. Shortly afterwards we headed down the ever-widening Firth of Clyde towards the open sea.

Our voyage would take us through the Mediterranean to the Suez Canal, then down the Red Sea, with calls at Port Sudan, to discharge a small part of our cargo, and Aden, to take on board bunkers. We would then proceed eastwards through the Indian Ocean and round past Dondra Head on the southern tip of Ceylon. From there our route would take us across the Bay of Bengal into the Andaman Sea, passing just to the north of Great Coco Island, which lies at the northern end of the Andaman Islands, before arriving off the Pilot Station at the mouth of the Rangoon River.

Once clear of the Clyde, and the continual distractions emanating from our head office at 95 Bothwell Street, we quickly shook down into our seagoing routine as we steamed steadily southwards. For this was an age when the master and chief engineer ran the ship without the doubtful benefit of the stream of e-mails, fax messages and satellite phone calls from head office which so be-devil, indeed demean, the mariner in today's world. In those days, communication with our head office was limited to routine voyage reports and stores requisitions, mailed at the end of the outward passage. A follow-up report and last minute store and spare gear request was mailed from the final bunkering port on the homeward passage: Aden in the case of the Burma ships, Las Palmas for the West African ships.

As 2nd engineer, I kept the 4 to 8 watch assisted by Davy Rice, the 5th engineer. The 8 to 12 watch was handled by the 4th engineer, Alfie Murdoch. The 3rd engineer, Ian Collins, took care of the 12 to 4, or 'graveyard', watch. The 3rd and 4th were each assisted by a junior engineer. Ian, Alfie and I had served our engineering apprenticeship with John G. Kincaid at Greenock at more or less the same time.

Later, Ian and I had served as junior engineers on the West African run on Paddy Henderson's *Kentung*, the chief engineer on *Kentung* at that time being none other than our present chief, Colin Kerr. So we all knew each other quite well.

Colin Kerr was quite a character – he addressed everybody as 'Sahib', presumably the result of his long service with the British India Company. He possessed a fund of homespun, highly-colourful expressions, most of which I had never heard before, not even when I was an apprentice in Kincaid's marine engine works, and have never heard since.

One of his more memorable expressions, which he frequently used when someone had upset him was, 'He should be shot with a ball of his own shit, Sahib; reeking-hot at point blank range!' My imagination never failed to boggle.

The bilge, ballast and general service pumps on the Paddy Henderson steamships and the early 'K' class ships were usually supplied by Dawson & Downie of Clydebank. These pumps were prone to causing the ships' engineers more than a little trouble once the valve chests on the steam end of the pump became worn. The bilge pump on *Kaladan* drove the 4th engineer and me crazy while we were anchored down off Port Harcourt in Nigeria one Christmas. After days of grinding and lapping the valve faces, the pump would start up very nicely; the 4th and I would congratulate each other on having overcome the valve chest problem by using just a little basic engineering knowledge. Then, after about half an hour, it would suddenly stop and refuse to start. The whole overhaul procedure would then be repeated, but with the same lack of success. The only way we could get the pump to operate was to have one of the firemen stand by it with instructions to belt the valve chest with a hammer every time it looked as if it was going to stop. By that method we pumped the bilges, twice daily, all the way back to the London River. The general consensus of opinion regarding these pumps was best expressed by Colin Kerr when he made the dramatic statement that, 'If all the Navies had been fitted with Dawson & Downie pumps in 1939 there would never have been a war. None of the bloody ships could have put to sea!'

During my *Kentung* days on the West African coast, we frequently carried deck passengers between the various coastal and river ports. They were accommodated on the foredeck where tents were rigged on the No. 2 and 3 hatches. One morning, as we steamed slowly up the River Benin towards the river port of Sapele, Ian Collins and I came up from the engine-room to have our morning coffee on the upper deck outside the engineers' accommodation. As we sat on the wooden deck, watching the tropical rain forest slide by on either side of the river, who should suddenly materialise before us but one of our deck passengers. From his flowing robes and distinctive head-gear he was a Hausa from up by Kano in northern Nigeria. He appeared to be a man with a grievance and subjected the two of us to a long, loud and impassioned harangue in a language that we strangers did not know. In the midst of this hullabaloo along the deck came Colin Kerr, curious no doubt to see what all the noise was about. Our Hausa deck passenger immediately recognised that here was a man of mature years, obviously someone of considerably more seniority than the two sweaty, oil-stained youths sitting drinking coffee on the deck, so he re-directed his complaints at our chief engineer. Colin Kerr listened carefully, indeed sympathetically, to what we took to be our friend's tale of woe and injustice, then announced gravely, 'There is no doubt about it old chap – your arse is most definitely out the window'.

*

7

On receipt of this verdict our Hausa tribesman shook the Chief's hand vigorously, then returned to his tented accommodation, confident in the knowledge that his grievances were being dealt with by a most senior person. Colin Kerr, on the other hand, went back to his cabin to work on his stores requisition sheets, or whatever, equally confident that his linguistic and diplomatic talents had triumphed once more. Ian and I looked at one another in silence, quickly finished our coffee and returned to the engine-room. Life was simpler down there.

However, to return to *Bhamo*, as we cleared the Irish Sea and entered the Atlantic the weather started to deteriorate and by the time we reached the Bay of Biscay we were in the grip of a full westerly gale. The ship rolled, pitched and corkscrewed in the confused seas. Down in the engine-room the main-engine revs alternatively raced and slowed down as the stern rose and fell in the sea-way. My memory is of being extremely tired at the end of each watch, the result of continually bracing myself against the constant, and at times fairly violent, movement of the ship as I did my rounds of the engine-room noting and logging the pressures and temperatures. This, of course, was long before the advent of automation and computerised print-outs of temperatures and pressures. In this sort of weather the ship was completely closed down and a musty, stale sort of atmosphere soon prevailed in the accommodation. We engineers never really saw the light of day; not that there was much light anyway, what with the short, northern, winter days and the foul weather conditions. We moved in a gloomy world from cabin to engine-room and back again with occasional visits to the dining saloon when we felt that our stomachs would be able to hold down a meal. The Burmese crew moved from their aft accommodation at the poop to the engine-room along the safety of the propeller-shaft tunnel. The engineers also made use of this subterranean route when inspecting the steering gear at the end of each watch. Thankfully, once clear of Cape Finisterre, the gale slowly began to abate and by the time we rounded Cape St Vincent, heading towards Gibraltar, conditions had greatly improved. In the Mediterranean the seas calmed down, the sun appeared with more than a hint of warmth in it, the weather doors and ports were opened and lovely fresh air flooded through the accommodation spaces. Suddenly, everyone was cheerful in this new bright and sunny world.

The machinery in our new ship gave no problems, with the exception of the fuel oil separators. These machines are designed to remove impurities in the form of water and solids from the fuel by means of centrifugal force. However, this results in a progressive build up of solid, sludge-like material round the rim of the separator bowl which has to be removed by stripping down and cleaning the separator. We were using a fairly unrefined heavy fuel and the rate of build up of solids in the separators was quite rapid, especially during conditions of heavy weather when movement of the ship tended to stir up the sludge at the bottom of the fuel tanks. To take care of this problem our separators were of the latest self-cleaning design, where, by opening and closing valves, the solids removed from the fuel were supposed to be washed away by high pressure water. Unfortunately, it did not work out like that in practice. On operating the self-cleaning controls, the internal discs moved apart to facilitate flushing but failed to re-seat; in addition, the bowl sealing rings were frequently damaged by partly removed solids. The end result was that we spent a large proportion of our time dismantling the separators, cleaning them, replacing damaged sealing rings and then reassembling them. It was always a race against time as we strove to avoid running out of clean fuel oil.

Davy Rice, on my watch, spent hours and hours on this task – all the way from the Med to Rangoon. Needless to say, he was not impressed with this latest, 'state of the art' design of separator.

'Och, see these automatic bluidy things – they're nae bluidy use tae onybody. I'd like fine to meet the big eedjit that invented them. Ah bet he wis a Celtic supporter.' was one of his more subdued vocal outbursts as he stood, steadying himself against the rolling and pitching motion of the ship, in the midst of a pile of separator discs, sealing rings, bearings and sundry other bits and pieces. Davy, of course, was a fervent supporter of Glasgow Rangers.

After six and a half lovely warm, but fresh, days in the Mediterranean we reached Port Said. One of the things I remember most about Port Said was the smell that wafted out on the warm land breeze to greet the ship as she came in from the open sea towards the Pilot Station. It was a strange, slightly spicy, but not altogether unpleasant, aroma that seemed to hint at the exotic magic of the Orient that lay ahead of us – or so it seemed to a lad brought up on a romantic diet of Kipling and Conrad. To those who lived and worked in Port Said it was probably more like a cocktail of cooking odours, bad drains, unwashed humanity and more than a whiff of raw sewage.

Another of my Port Said memories is the 'bum-boat' wallahs who used to scramble on board, selling everything from dirty postcards to bottles of fake Scotch whisky. They were without doubt the world's finest imitators of the Scottish, and in particular the Glasgow, accent. Their skills had been honed to perfection by years of conversing with Scottish ships' engineers, not to mention the thousands of Scottish troops (51st Highland Division and others) with whom they no doubt carried out 'business negotiations' circa 1941–43.

After picking up the pilot, *Bhamo* made fast to buoys in the harbour at around 8.00p.m. and waited to join the next south-bound convoy. Once I had handed over the watch to the 4th engineer I came up out of the engine-room and stood on deck, surveying the scene. We were moored more or less opposite two very familiar landmarks – the prominent neon sign advertising the well-known waterfront department store, Simon Arzt, and the equally prominent neon sign for Johnny Walker whisky. Davy Rice joined me just as a bum-boat wallah scrambled over the ship's rail almost in front of us. Looking at Davy he said, 'You wanta haircut, meester – very good haircut, very cheap price.'

'Naw,' said Davy. That's all he said – just that one word. But it was enough for the bum-boat wallah.

'Aaw, whit's wrang Jock? Whit wey are ye no wantin a haircut the day?' was the Egyptian's immediate reply in a voice that was straight from Sauchiehall Street.

Davy was so astonished his jaw nearly hit the deck. 'How does he ken I'm frae Glesgie?' he gasped in total amazement.

'Ah, it's just one of their many talents, Davy,' I told him. 'I hope your cabin door is locked,' I added. 'They have other, not so friendly, talents.'

Early next morning we joined the convoy and steamed slowly through the Suez Canal, past Qantara, Ismailia and the Bitter Lakes, to reach Port Taufiq at the southern end by late afternoon. My memories of Canal transits are nearly all of hours spent on the floor plates, drinking gallons of coffee while adjusting the main-engine revolutions up and down to enable those on the bridge to keep us equidistant from the ships ahead and astern of us. Up top, the scenery was quite boring – just miles and miles of sand, stretching away on either side of the ship.

After clearing the southern end of the Canal at Port Taufiq we steamed out into the Gulf of Suez, then down the Red Sea to Port Sudan, where we discharged about 1,000 tons of cargo. There, we engineers worked tropical hours: from 6.00a.m. to 1.00p.m. This enabled us – apart from the duty engineer, of course – to spend our afternoons at the Red Sea Hotel, drinking ice-cold beer and swimming in the hotel pool. Despite having known Alfie Murdoch, our 4th engineer, for several years during our apprenticeship days at Greenock, I was surprised to discover that he could not swim. Ian Collins and I took it upon ourselves to rectify this unsatisfactory state of affairs. 'Don't worry, Alfie,' we told him. 'We will have you swimming before we leave Port Sudan – there's nothing to it.' Alas, poor Alfie had all the inherent buoyancy of a ton of bricks and, in the end, we had to admit defeat before he drowned himself or, what was much more likely, all three of us.

From Port Sudan it was down the Red Sea to Aden of the Barren Rocks, where we took on approximately 500 tons of fuel oil. I always hated bunkering calls at Aden or Las Palmas. Invariably, we would arrive towards the end of my 4.00p.m. to 8.00p.m. watch. Supervision of the bunkering operation would then occupy me for the best part of eight or nine hours, following which we would prepare for sea and depart some time during my 4.00a.m. to 8.00a.m. watch. The end of the watch would be followed by a very hearty breakfast, after which I would climb into my bunk and sleep the sleep of the just – and the very weary.

The distance from Aden to the Rangoon Pilot Station is 3,310 nautical miles and, at our service speed of 14 knots, took us ten days. These were invariably very pleasant days. Generally, the sea was like a millpond and, as we sunbathed on deck during our off-duty periods, we could watch schools of dolphins and flying fish cavorting around the ship. Occasionally, manta rays appeared, rising dramatically from the water and splashing back down again. Apart from watching the sea-life, deck quoits and deck tennis were popular ways of passing the time. These games were usually accompanied by raucous barracking from the spectators – and, indeed, one's opponents. In the cricket world I believe this is now termed 'sledging'. Our players, who were generally of a most robust mental outlook, were not affected in the slightest way.

In *Bhamo* I made my first visit to the Far East. After picking up our pilot in the evening we moved up river and anchored off Syriam, some five miles downstream from the port of Rangoon. At about 6.00a.m. next morning I came up on deck to see what was going on. Sitting on the No. 4 hatch cover, drinking my early morning tea, I looked up river towards Rangoon. A low, white, tropical mist lay over the city but, sticking up out of the mist, sparkling in the morning sun, was the huge golden spire, or stupa, of the Shwedagon Pagoda. Slowly, the mist cleared away and the city of Rangoon, complete with what appeared to be dozens of gleaming, golden pagodas, emerged into the tropical morning. To me this was the start of Kipling's 'Mandalay'. All the magic of the Orient seemed to lie up the mysterious Irrawaddy, where he said that '. . . the dawn comes up like thunder outer China 'crost the bay!'

At 10.00a.m. we weighed anchor and steamed up river to berth alongside the Sule Pagoda wharf. Hatches were opened and, shortly afterwards, discharge of our cargo commenced. In the engine-room we changed over from sea to port conditions and then took the rest of the day off. Our journey had taken just over one month.

Our days in Rangoon were very pleasant. In the engine-room we worked our tropical hour routine, turning to at 6.00a.m. and stopping at 1.00p.m., or sometimes an hour later if we were busy. With the aid of advice received from the manufacturers we now set about dealing with the problems we had experienced with the fuel oil separators during the outward voyage. Colin Kerr had fired off a snotty complaint to them, via our head office, from Port Said, and their response was waiting when we arrived at Rangoon.

In addition, we commenced overhauling one of the main-engine cylinder units. This involved withdrawing the top and bottom pistons, cleaning the pistons, piston rings and cylinder liner; also testing the cylinder lubricators. The running hours on the engine did not, as yet, require the removal and cleaning of any of the pistons, but we thought it a good idea now to try out our, as yet untried, tools and overhauling gear. It was a good thing we did, because the spanner for the bottom piston/crosshead nut did not fit properly between the upper fork of the connecting rod. This spanner was a fairly massive piece of equipment that really required two of us to handle. I think it took me several days of grinding away at the outside surface before we were able to use it. In the process we used up all our supply of grinding stones and ended up with the workshop floor knee deep in iron filings. While all this was going on, Ian Collins was busy carrying out routine maintenance on the Ruston diesel generators and checking the main-engine fuel injection valves. Alfie Murdoch was occupied overhauling the air compressors and various pumps. The electrician was fully employed cleaning and adjusting generators, motors and starter boxes; and also keeping a careful eye on the cargo winches to make sure that the discharge of cargo was not interrupted by any failures of winch motors or controllers. This was the first real test of the cargo winches, the loading of cargo at Glasgow having been carried out by shore cranes.

The Burmese engine-room ratings provided considerable assistance to the engineers during maintenance work on the machinery. The average Burman appears to have a natural mechanical talent, an everyday example being the number of rather ancient vehicles that could be seen plying the streets of Rangoon. The maintenance of these vehicles, many of which appeared to date from World War II, cannot have been easy, with most, if not all, replacement parts having to be manufactured in local back-street workshops.

After watching, and assisting, Davy Rice and others battling with our recalcitrant fuel oil separators on the outward voyage, two of our Burmese ratings were able more or less to take over the job of stripping them down, cleaning and then reassembling them on the homeward passage.

In our free time we did a bit of sight-seeing and a 'must', of course, was a visit to the magnificent Shwedagon Pagoda. This massive pagoda, the greatest in all of Burma, if not the world, dominates both Rangoon and the surrounding flat delta country. It is thought to be approximately 2,500 years old and is the holiest shrine in Burma. On Sundays the British Sailors' Society padre took us out to the Inya Lake, where we spent most of the day pottering about in a couple of sailing-dinghies. The lake was a beautiful, peaceful place, a pleasant change from the hustle and bustle of the city and the docks. Sometimes, after dinner, we would stroll out of the docks and along to the Strand Hotel for a quiet drink in the bar. The Strand is one of three famous old hotels in the Far East, originally owned by four very astute Armenians – the Sarkies brothers. The other two being the E & O in Penang and, most famous of all, Raffles in Singapore.

Out greatest off-duty interest was probably football and even before we arrived at Rangoon we had been busy organising the ship's team. With only seventeen British officers we did not have many players to choose from. The master, chief engineer and chief steward were regarded as being rather over the hill as far as playing football was concerned, so that left fourteen potential players. Unfortunately, there had to be at least one engineer and one deck officer on board at all times. That reduced the pool to twelve – eleven players and one reserve! Colin Kerr was very keen, his brother-in-law was Scott Symon, the manager of Glasgow Rangers, so with that sort of impeccable football background he appointed himself team manager.

We played several games against other ships' teams, including Bibby Line's *Warwickshire*, and the local St Paul High School who were very good. I don't think we ever won any games, but we drew with *Warwickshire*. There was no way we were going to let a bunch of 'scousers' beat us. However it was all great fun and the exploits of the football team gave everybody on the ship something to talk and argue about – especially Colin Kerr, who carried out a thorough analysis of the team's performance after each game. His usual conclusion? We were rubbish! He was probably correct.

After fourteen days alongside the Sule Pagoda wharf, discharge of our general cargo was completed and we moved out into the stream, where we were secured to buoys fore and aft for a couple of days, waiting to start loading our homeward cargo. The first part of cargo, 1,500 tons of bagged rice, turned out to be available at Bassein, a port lying on the western side of the Irrawaddy delta, some 60 miles up the Bassein River and 200 miles from Rangoon. Leaving Rangoon at 8.00a.m. one morning we anchored at the mouth of the Bassein River in the evening, then steamed up to Bassein after daybreak the following morning. It took three days to load the bagged rice – then it was back to our double-mooring in the Rangoon River, where we spent another fourteen or so days loading teak logs, sawn timber and more bagged rice. Two days before departure for home we moved downstream to cross the bar and anchored off Syriam as, at her fully loaded draught, *Bhamo* would not have been able to cross the bar without taking the bottom.

An interesting incident occurred while we were anchored off Syriam. I was working on a problem with the chilled-water line, located in the vegetable room, when suddenly two Burmese customs officers appeared. One of them had a copy of our crew list and, pointing at our engine-room cassab's (storekeeper's) name, he said, 'We have information that this man bought hashish at a shop in Rangoon yesterday.'

While I was digesting this startling information, the two customs officers started rummaging around the vegetable room. Within a few minutes they produced two cardboard packets, each about $10 \times 5 \times 5$ inches in size. On being opened, they were found to contain a tobacco-like substance that the two customs men identified as the hashish the cassab had purchased the day before. I was amazed. There I had been, quietly working away on a defective water line connection, unaware that about £10,000 worth of illegal drugs was parked under my nose. For a horrible moment I thought I might be arrested as an accessory. After all, the cassab worked directly for me and the drugs had been discovered right beside where I was working. What more circumstantial evidence did they need? However, the customs men were not in the slightest interested in me – nor even the cassab, apparently. They had sucessfully recovered the drugs and that seemed to be all that they were interested in. They just piled into their launch and sped back to Rangoon, mission accomplished.

I have always been puzzled about that incident at Syriam. Presumably, someone in our crew must have informed the Customs. Otherwise, how would they have known where to find the hashish. They went straight to the vegetable room, never looked anywhere else. The fact that they were able to locate the two packets almost immediately among the piles of provisions in the vegetable room would appear to indicate that they had been told exactly where the hashish had been hidden. This information could only have come from someone on the ship. The cassab, of course, who was a most efficient member of the engine-room staff, denied all knowledge of the affair. He may well have been correct in his denial – anyone could have given the cassab's name to the Customs; on the other hand, it might have been him. I

17

wondered what the customs men did with the recovered drugs – sold them, probably.

With the final parcels of cargo loaded, hatches and derricks secured and our fresh water tanks topped up, we slipped down the Rangoon River, homeward bound.

The machinery performed well all the way home. The problems previously experienced with the fuel oil separators had largely been overcome, as a result of the modifications carried out at Rangoon, and we now only had to strip them down and clean them manually about every six or seven days, whereas, on the outward voyage, we had never gone more than two days without having to strip down one or both of them.

The homeward passage progressed without incident. We stopped again at Aden for bunkers and then proceeded through the Red Sea and Suez Canal. All the time, everyone, both on deck and in the engine-room, was busy cleaning up the grime and debris that had accumulated during our stay at Rangoon. In the engine-room, paint-work was washed down and, where necessary, repainted. Everything that could be polished was polished until it gleamed. Colin Kerr, surveying the result as we neared home, was moved to remark in his own inimitable manner, 'She's looking great, Sahib – shining like a tanner in a nigger's arse!'

Despite the excitement of visiting the East and other faraway lands, there was always a very special joy in coming home and catching your first sight of the Scottish hills. It was on a bright, cold, almost frosty morning in April 1958 that *Bhamo* arrived in the Firth of Clyde on completion of her maiden voyage. Nowadays, few Scots are able to experience the thrill of arriving back home in Scotland by sea. Transatlantic liners no longer call at the Tail of the Bank to drop off and pick up passengers; now everyone travels by air and, with few exceptions, one airport is just like any

18

other. By comparison, the approach to the Clyde from the open sea is a really wonderful, magic experience.

We were due to pick up the pilot off Gourock at about noon so, after finishing my morning watch at 8.00a.m., I raced up the engine-room ladders to see where we were. We had just entered the Firth and, to starboard, the symmetrical granite cone of Ailsa Craig, or 'Paddy's Milestone' as we knew it, rose out of the sparkling, blue water. Later, the Isle of Arran came abeam on the port side, with the tops of Goatfell, Cir Mhor and Casteal Abhail sparkling in a covering of late-season snow. Over to starboard were the Heads of Ayr and the rolling green fields of the Ayrshire country-side – the heartland of Robert Burns, Jean Armour, Souter Johnny and Tam O' Shanter. As we steamed on, the Island of Bute came into view over to port, giving us a glimpse of Rothesay Bay and, just beyond, the entrance to the fabled Kyles of Bute. Ahead, the snow-covered sierra-like tops of the Arrocher Alps – the famous Cobbler, Ben Ime, Ben Vane and Ben Vorlich – formed an impressive backdrop to the lower, russet-coloured Argyll hills round Loch Long, Loch Goil and the Gareloch. Soon the Cloch Lighthouse drew abeam and we rounded to starboard, passing between Kilgreggan and Gourock, where we picked up our pilot, before anchoring at the Tail of the Bank to await the tide. As we slowed to pick up the pilot, I caught sight of something white fluttering from an upstairs window of a Cardwell Bay house and smiled to myself; my mother had spotted us. She would have been down at the Pilot Office on the pier-head the previous day to find out exactly when Paddy's *Bhamo* was due in the Clyde.

After a couple of hours we weighed anchor and headed into the river channel, past the Greenock and Port Glasgow shipyards; as we passed Lithgow's yard some of the workers gave us a wave as they recognised one of their own ships coming home. On we went: up the river, past Langbank, Dumbarton Rock, the Kilpatrick hills and Erskine, towards the city of Glasgow. By 4.00p.m. we were snug alongside the 'Burma Boats' berth' at Plantation Quay and the Bridge had rung down 'Finished with Engines' for the last time on the voyage. Down in the engine-room we changed over to harbour conditions as quickly as we could. Everything went smoothly and, looking at the clock, I began to think to myself that, barring some unforeseen mishap, I might just be able to catch the 6.00p.m. Glasgow to Gourock bus that evening, down home to Cardwell Bay.

2

Port Out, Starboard Home

The word 'posh' is alleged to be an acronym for the term 'Port Out, Starboard Home'. In the days of Empire, discerning colonial and military officers and other globe-trotting passengers travelling to India and the Far East are said to have booked port-side cabins for the passage out and starboard-side cabins when returning. The idea being that port-side cabins would be cooler going east, as the sun would be on the starboard side – the reverse being the case when steaming west on the homeward voyage. Presumably, this sort of cabin allocation involved a premium charge, so only those who could afford to do so travelled in this superior, or 'posh', way.

I had left Paddy Henderson's *Bhamo* in December 1958 and went off to sit my chief engineer's examination. Once that hurdle was over I joined the P & O Company and spent two years in their steam-turbine driven ships, sailing to India, Hong Kong and Australia. After initially spending about six weeks on relieving duties at Tilbury Docks, I was appointed as Junior 2nd engineer on *Carthage*, a thirty-year-old passenger ship on the London to Hong Kong run, a service maintained by two other similar ships, *Corfu* and *Canton*

The three ships operated a regular service from London, via Southampton, to Hong Kong with intermediate calls at Bombay, Colombo, Penang and Singapore. On completion of cargo loading and voyage repairs at the Royal Docks in London the ships would depart in the afternoon for the passenger terminal at Southampton, normally berthing at about 8.00a.m. the following morning. The passengers, usually three hundred or so, arrived on the boat train from Victoria station in the middle of the morning and we would be off down the Solent in the late afternoon.

My cabin on *Carthage* was on the starboard side of the ship and it was certainly warm enough as we headed east towards India. I wondered if it would be any cooler on the way back. Time would tell.

As Junior 2nd engineer I kept the 4 to 8 watch and was assisted by two other engineers, Don Large from Taunton who stood the watch in the boiler-room, and Billy Waring from Manchester who looked after the engine-room. Waring was quite a character – of a somewhat rotund appearance, he was popularly referred to as 'Pudding' by all, except Jock Chivas, our chief engineer, who, on the few occasions when he condescended to address him, simply referred to him as 'Waring'. Their first meeting, as far as I can remember, was the morning after we had sailed from Southampton. Waring was on his way up the engine-room ladders to make sure the 8 to 12 watch were awake and ready to come on duty. Half way up he met Jock Chivas who was on his way down to commence his morning rounds. Waring's description of their meeting went something like this. ''Ere, oo's that fat fellah just coom down engine-room? Ah was 'alfway oop to call watch like, when eeh appears on t' ladder. "Get down, get down, boy," eeh shouts and ah had to coom right back down t' bottom. Ooh does eeh think eeh is?'

'That', I explained, 'is the Lord of the Manor – Jock Chivas, our chief engineer. For God's sake, don't get in his way again or we will all suffer'. Even at this early stage of our acquaintanceship I had discovered that, even in a good mood, Jock was not an easy man to deal with. In a bad mood, he could be very difficult indeed. He had a disconcerting habit of firing questions at you, then, as you tried to reply, shouting, 'What! What!' As a consequence he was sometimes referred to as 'Old What What'.

During my spell on the Dock Staff at Tilbury I had found that chief engineers on P & O passenger ships were treated with considerable deference. At 6.00p.m. one afternoon on *Strathmore*, while several of us spare 2nd engineers were having a pre-dinner gin, a large gentleman in shore attire entered the cabin. Immediately everyone stood up. After a few words our visitor departed and we all sat down. 'Who was that?' I enquired. 'Oh that was the chief engineer,' was the reply. Obviously, life in the P & O was going to be a bit different from the fairly casual, relaxed atmosphere that I had been accustomed to in Paddy Henderson. Actually, in practice, it was not all that much different really. But the chief engineer, no matter who he was, was always treated with considerable respect.

The size of the engine-room staff on *Carthage* was much larger than anything I had previously been used to. There were nine watch-keeping engineers (three per watch), two refrigerating engineers and two electrical engineers. We also carried a boilermaker, one Steve Wilson, a massive, extremely muscular young man who gave the impression that he could, with ease, bend boiler tubes with his bare hands. All the P & O steam-turbine passenger ships at that time carried a qualified boilermaker, presumably to ensure that any in-service boiler problems could be dealt with on the spot. Apart from checking the condition of our six boilers when they were shut down during our stay in Hong Kong, one of our boilermaker's tasks was the management of the fuel oil system, including the supervision of bunkering operations at Aden and other ports. So, to my great delight, for the first time in several years I had no part to play during bunkering.

23

The instrumentation in the old *Carthage* was very basic and of somewhat questionable accuracy. You most definitely steamed the boilers and operated the turbines by 'the seat of your pants' rather than a careful monitoring of the instruments. In addition, the engine-room was extremely hot; it was by far the hottest engine-room I ever experienced. In the tropics the heat in the generator flat, which housed three steam-turbine-driven generators, was simply unbelievable. We used to joke that more steam escaped through the glands into the engine-room atmosphere than passed through the turbines. Waring graphically described the engine-room one day during a south-bound Suez Canal transit. Arriving beside me on the control platform, having just finished an inspection of the machinery, he wiped his perspiring face with a sweat-rag and said, 'Eeh, it's joost like summat outa Ming Dysentery.' We had almost reached the southern end of the canal before I worked out what was wrong with that statement.

These days, some forty years on, I often watch the Antiques Road Show on a Sunday evening on television. Occasionally, when I hear one of the experts holding forth about some ancient oriental vase, I close my eyes and am a few thousand miles away, back in the engine-room of the old *Carthage*. My sweat-soaked overalls are stuck to my back and I am listening to Billy Waring giving me the benefit of his candid opinion on the whole antiquated set-up.

Ever since his inauspicious first meeting with the chief engineer on an upper engine-room ladder, Billy kept out of his way. After all, there was no point in looking for trouble. All was well until we arrived at Hong Kong for our five-day scheduled stop-over alongside the wharf at Kowloon. On the second night we were there, Billy went for a run ashore with several of the engineers. On returning to the ship at around midnight he decided to go for a swim and cool off.

Stripping down to his underpants, he dived from one of the gun-port doors on the working deck into the harbour. His dive and the subsequent splash was observed by the 3rd officer, who had charge of the deck that night; he called the duty engineer, presumably on the logical basis that the engineers should look after one another. The engineer in question was the 3rd engineer, Stan Smith, a suave Londoner who, at the time, had possibly had a couple of gins too many. On being alerted, he rushed up to the gun-port door and, after divesting himself of most of his uniform, dived into the harbour to the rescue. Unfortunately, Stan was not really a very strong swimmer, whereas Billy Waring, like many overweight people, was. Anyway, Stan eventually spluttered up to Billy, who was swimming about quite nicely feeling suitably cool and refreshed. 'What's oop Stan?' asked our aquatic star.

'What's up!' gasped Stan. 'I've come to save you "Pudding" – that's what's bloody well up!'

Meanwhile, the 3rd officer, on observing Stan floundering about and fearing that the rescue operation was not going as well as it might, had notified the Harbour Police and within a few minutes our two swimmers were picked up and returned to the ship. Unfortunately, the intervention of the police resulted in an official report being submitted, and this was brought to the attention of the chief engineer next morning. The outcome was that I was detailed to bring our two swimmers along to his office to explain what it was all about.

Waring started off by saying, 'Ah was feeling a bit 'ot like, so ah thought ah would 'ave a little swim; ah mean nobody said nowt when ah had a swim at Bombay'.

Stan Smith and I exchanged a startled glance. This was the first we had heard of him, or anybody else for that matter, swimming in Bombay harbour.

Jock Chivas was more than startled. The thought of any

of his engineers swimming in the harbour at Bombay was simply beyond his comprehension.

'Bombay!' he shouted. 'Bombay! You went for a swim at Bombay? But it's filthy, man – the water's filthy. I can't believe it. What have you got to say? What! What! Speak up, man, speak up! What! What! Oh, never mind ... Walker, get him out of here before we all catch some terrible disease. You better take him along to the doctor while you're at it; he's probably got something highly infectious. Go on – get him out of here! What! What!'

I could hear him still ranting on as we fled along the alleyway, trying our best to suppress our laughter. I am sure Billy Waring never swam in Bombay harbour – the mind boggles at the very thought. It was just a ploy to throw Jock Chivas off his stride. Mind you, with Waring you could never be absolutely sure. Anyway, whatever the truth of the matter, it certainly brought the 'Court Martial' to an abrupt and complete stop. Stan was rather disappointed at the outcome. I think he had hoped for a 'mention in dispatches', or whatever, on account of his spectacular, albeit unsuccessful, rescue attempt.

Another interesting character was a South African who joined in London when I was up in Scotland for a few days' leave between voyages. On returning to the ship I was greeted by a large, cheerful fellow who introduced himself as Pete Kruger, a new junior engineer. 'I served my apprenticeship at Johannesburg with Stewart & Lloyds but I've been over here almost a year now,' he told me.

'What have you been doing over here. Where were you working before you joined P & O?' I asked him.

'Oh, nothing to do with engineering. I've been working in show business – had a part as an extra in a couple of films and an all-singing-and-dancing part in a musical,' came his astonishing reply, which rendered me totally speechless.

A large percentage of our passengers were government

26

servants, such as Embassy and High Commission staff coming and going to appointments in India, Malaya, Singapore, Hong Kong and China. There were always numerous Gurkha Officers travelling on leave or returning to rejoin their regiments in Malaya, Singapore or Hong Kong. We also always had a substantial number of passengers from the Hong Kong Police, again either travelling to the UK on leave or returning to Hong Kong.

There was always a good rapport between the ship's officers and our Hong Kong Police passengers. On one voyage this led to our 2nd officer, Cecil Smillie, and one of the policemen, in a fit of misplaced enthusiasm, arranging for the ship to play a game of rugby against a police team after we had berthed at Kowloon. After Cecil's initial excitement had died down reality set in, and a bit of difficulty was experienced finding fifteen crew members who were fit enough and knew enough about the game to be able to form a team. Only Cecil, the ship's doctor, about three or four of the junior officers, including Pete Kruger and myself, had played before. However, by a process of eliminating the obviously over-age, over-weight and totally unfit, we managed to find a sufficient number of good men and true to make up the ship's fifteen. The fact that some of them had to be bribed by the promise of a great party in the police mess after the game was neither here nor there. Our radio officer, Bob Hargreaves, was a fund of sporting information. Every Sunday he produced a special sports edition of the ship's daily newsletter. Nobody on the ship knew more about rugby than Bob; his rugby stories, especially when encouraged with a couple of double gins, were legion. Such knowledge could not be overlooked so he went straight into the team, despite concerns over his fitness and weight. Anyway, we needed his weight in the scrum. Billy Waring and Steve Wilson, the boilermaker, were picked for much the same reason.

What Cecil and his police friend failed to tell us at this stage was that the Hong Kong Police rugby team had not lost a game for some considerable time. In the last couple of years they had beaten all the British Army, Airforce and Navy teams in the colony; only the New Zealand cruiser, HMNZS *Black Prince*, had managed to draw with them. I suppose they reckoned that what we didn't know wouldn't do us any harm. Fat chance!

All fifteen of us collected at the Boundary Street Police Station sports ground at 5.00p.m. one afternoon and, shortly afterwards, the game was on. Our opponents were not actually the Police 1st XV but even the reserve, or 2nd XV were formidable – 'frightening' might be a more accurate description. Within ten or twelve minutes our opponents had run-in four tries and, rather patronisingly, announced that they would not be bothering with any conversions. To say that my memories of that game are a bit blurred would not be an exaggeration. On the few occasions when I managed to get hold of the ball I was either promptly smashed into the ground or hurled into touch. I can remember Bob Hargreaves, our sporting correspondent, emerging from a scrummage with his shirt half-torn off and his nose bleeding.

'How are you doing, Bob?' I asked.

'Bloody hell!' was all he could gasp at me as he staggered off.

But you can't beat a bit of hands-on experience, can you? Bob's sport reports in the Sunday newsletter would now have a ring of total authenticity about them.

My clearest memory of the afternoon is at about half time when Cecil, the doctor, Pete Kruger, Steve Wilson our muscular boilermaker and several others, managed to get the ball across our opponents' line for a try. Cecil rose to the occasion and announced with great panache that, just like the police, we would forgo the conversion. I can't now recall exactly what the final score was but the police must have scored at least ten or twelve tries, possibly more, to our one and only – and that was without trying too hard.

These days when I read, or hear on the television, about some sporting 'mis-match' or other, my thoughts slip back to those 60 minutes of torture in the sun at the Boundary Street Police Station sports ground at Kowloon. That was the Mother of all 'mis-matches!' But never mind. The Hong Kong Police treated us most royally in their mess after the match; they could not have been more hospitable. Goodness knows when we all straggled back to the ship at the end of what had been for all of us a most memorable run ashore. We even had the bruises and sore heads to prove it!

Next day, feeling just a bit fragile, I was quietly going about my business in the engine-room when out of the blue I received a message from Cecil Smillie. Could I put together a soccer team to play one from RAF Kai Tak at 5.00p.m. that afternoon? To say that playing football was the last thing I was thinking about that day would have been an understatement. But someone, probably Cecil, had promised that the ship would have a team out at the 'Missions to Seamen' sports ground in Kowloon at the appointed hour. So there was nothing for it but to find eleven players and get out there by about 4.30p.m. in time for a bit of a warm up before the opposition arrived. It was not quite so difficult rounding up the necessary eleven bodies, although several of the previous day's rugby players were unavailable owing to their having been 'knocked about a bit'. The boilermaker was also unavailable because of a problem with one of our main boilers and the scheduled arrival of some 300 tons of fuel oil at about kick-off time. He did promise to come along later and give us a bit of support during the second half.

Compared with the previous day's rugby game this match turned out to be fairly uneventful and not quite so one-sided, although in the end the RAF team beat us by a couple of goals. The only little bit of drama occurred after the game when we invited the RAF lads to join us for a beer or two in the bar of the Seamen's Club. As luck would have it, just as the first of the RAF players reached the bar and ordered up a round of San Miguel beers they ran into half a dozen or so sailors from a German ship who gave the impression that they were rather upset to find that non-seafarers were being served in what they apparently thought was exclusively a seaman's bar. They were obviously somewhat unhappy about the situation and I could hear words like 'Donner und Blitzen', 'Englander' and 'Luftwaffe' being bandied about in a fairly aggressive manner. It fell to me, as the senior Merchant Navy officer present, to explain to the Germans that the RAF football players were in fact our special guests and they were not to be harassed. This proved to be not so simple as one would have thought and my explanation was continually interrupted by such Teutonic remarks as, 'Vy are dese people in ze bar? It is seamen's bar, ja. Zey in dere own bar should be, is not so.' . . . and more of the same ilk.

I was beginning to get a little concerned by their attitude, and my failure to defuse the situation diplomatically, when suddenly they changed their tone and one of them said, 'Ah, so it is ze futball you are playing. Zat is very goot, ja. Ve are some times playing ze futball also.'

This sudden change of manner was not because they had at last cottoned on to what I had been trying to tell them, but rather that they had just caught sight over my shoulder of the awesome shape and size of Steve, our boilermaker, who had, as he had promised, turned up to support his ship's team. His arrival on the scene was most timely, for even a fleeting glimpse of Steve must have made it obvious to our German 'friends' that this was a man who should not be taken lightly. A black eye and some other facial bruises, courtesy of the Hong Kong Police rugby team, only served to emphasise a somewhat fearsome appearance. Mind you, like many big strong lads, Steve wouldn't normally have punched a hole in a wet newspaper. He was really a big, gentle giant but I didn't bother to explain that to the Germans.

All was sweetness and light after that. Which just goes to prove that, as in the world of international affairs, there is nothing to beat a bit of diplomacy, backed up by a bit of muscle, to get your point across.

By the time we left Hong Kong, a couple of days later, those of us who had done battle against both the police and the RAF were more than a little exhausted and looking forward to a return to the comparatively peaceful routine of shipboard watch-keeping as we steamed homeward down the South China Sea, through the Malacca Strait and across the Indian Ocean towards Aden and then Suez.

My favourite eastern port of call was Penang, where *Carthage* called on both outward and homeward passages. This island, often referred to as the 'Pearl of the Orient', lies some three miles off the north-west coast of the Malayan Peninsular at the north-western end of the Malacca Strait on latitude 5 degrees 20 minutes north and is approximately 15 miles long, north to south, and 8 miles wide. The central spine consists of a range of thickly for-ested hills, rising to a height of almost 2,750ft at the northern end near Georgetown, the capital.

My first view of Penang was just after daybreak one morning when we were lying off the north coast waiting for the pilot to take us into Georgetown. I had nipped up from the engine-room manoeuvering platform to have a quick look through one of the gun-port doors, which were usually secured in the open position in the tropics to aid ventilation, to see where we were. The view was simply beautiful: we were probably two miles from a lush, green island with a band of low, white, tropical cloud lying over it. Rising above the white cloud into the clear blue sky were the tops of the central mountain range. To me it was exactly what I had imagined a tropical island would be like – only more so.

We berthed alongside Swettenham Pier, Georgetown, shortly after 7.30a.m. and, as soon as our watch finished at 8.00a.m., my two watchkeeping colleagues and I washed and had a quick breakfast. Then with the 'Childy Ho' (Children's Hostess) in tow, we grabbed a taxi and set out from Georgetown along the north coast of the island, through Tanjung Bungah, past the Ghurka leave-centre at Tanjung Huma, to arrive at the Lone Pine hotel on the beach at Batu Ferringhi. There we spent the morning swimming and sunbathing before enjoying an open-air buffet lunch under the casuarina trees that fringed the beach. It was just delightful; it made putting up with all the ship's problems, such as the oppressively hot and antiquated engine-room and our irascible chief engineer, almost worthwhile. Then, after lunch, it was back to Georgetown and back on watch at 4.00p.m., refreshed in both mind and body.

We sailed from Penang at 6.30p.m., heading westwards towards the setting sun and our next port, Colombo. According to the laws of 'POSH' my starboard side cabin should now have felt a little cooler. I am afraid I never noticed any difference; so 'Port Out, Starboard Home' may well be nothing more than an old colonial myth.

In those days, almost 45 years ago, the Lone Pine was the only hotel on the two-mile stretch of the Batu Ferringhi beach. At one end was the completely self-contained Ghurka leave-centre, at the other the little Malay village of Kampong Batu Ferringhi. It was an idyllic location. Inevitably, Penang was 'discovered' by the holiday tour operators and by the late 1970s tourists were arriving in Penang in large numbers. Many were attracted to the lovely sandy beaches on the north coast and additional hotels sprang up along the Batu Ferringhi beach-front. Where once there was only the Lone Pine, shaded by its casuarina trees, there are now eight other hotels, plus a whole host of restaurants, batik shops, tailors, money-changers and stalls selling all sorts of copied designer clothes, watches, CDs and DVDs. But, despite the over-development, Penang still remains one of my favourite spots in the whole world. In recent years I have stayed several times at the Bayview Beach Resort at the western end of Batu Ferringhi. To stroll along the beach in the early morning, or just as the sun dips into the Andaman Sea, is, for me, quite magical. Sometimes, as I gaze out to sea from the beach, I see in my mind's eye a rather old-fashioned looking, but familiar, white-hulled P & O passenger ship materialise from around the headland; and above the hum of the turbines I can clearly hear Billy Waring giving me, in his distinctive Mancunian tones, his forthright views on something or other that has upset him, and Jock Chivas shouting, 'What! What! Speak up, man. What!' Then they have gone and I am left with just the sand, the sea and the sun and some very happy memories.

3

Liberty Ships

The Singapore firm of Ritchie & Bisset, Consulting Engineers and Marine Surveyors, which I joined after leaving the Merchant Navy, traced its origins back to 1866. In that year, almost a century and a half ago, Charles Fittock, a ship's carpenter, reached the port as one of the survivors of a shipwrecked British sailing ship. In those days of wooden sailing ships the carpenter was a very important member of the ship's company; in Fittock's case the captain regarded him as not just a carpenter but a master shipwright, the term used at that time to describe the fine craftsmen who built and repaired wooden ships. While awaiting onward transportation from Singapore, Fittock was approached by the representatives of several insurance companies who, because they had sustained heavy losses over the previous few years, had been considering the possibility of appointing someone who had sufficient technical knowledge to advise them on the claims being submitted by the large number of wooden-hulled sailing vessels that traded in and out of Singapore. After giving the matter careful consideration Fittock agreed to give it a trial. After only a few months he proved his worth and was soon in business for himself as *Chas. Fittock, Shipwright Surveyor*.

His business spread and he eventually became the Singapore representative of Lloyd's Register of Shipping. He remained in Singapore for more than 40 years before selling out to one Thomas Adam, then Chief Engineer on the Blue Funnel liner, *Charon*. The firm then became Fittock & Adam, Shipwrights, Engineers & Marine Surveyors. In August 1913, a Shell tanker struck a reef near Manila and Adam was appointed by the Salvage Association of London to attend the vessel and ascertain the prospects of successful salvage. Not only did Adam supervise the refloating operation, but was able to steam the casualty into Hong Kong under her own power.

In 1914 Adam invited Fred Ritchie, who was then Government Surveyor of Ships at Penang, to join him. Unfortunately, Adam, whose health had been deteriorating, died three weeks later and Ritchie took over the business. Ritchie, then aged 28, was full of energy and talent. He quickly built up the business and in 1916 secured as his partner David Bisset; the firm then became known as Ritchie & Bisset, Consulting Engineers and Marine Surveyors. They worked for all the major insurance companies and represented nearly all the ship classification societies, with the exception of Lloyd's Register who by this time had their own surveyor in Singapore. In addition they carried out consulting work for Straits Steamship Company and designed many of their vessels; the two most notable perhaps were *Kedah* built by Vickers Ltd at Barrow-in-Furness in 1927 and *Rajah Brooke* built by the Caledon Shipbuilding Company at Dundee in 1947.

Before the war *Kedah* was the pride of the Straits fleet, operating a fast passenger service between Singapore and Penang. Her twin-screw, high-pressure steam-turbine machinery gave a service speed of 20 knots, making her very suitable for conversion to the naval rôle of fast patrol-vessel in 1939. On the last organised convoy out of Singapore on 13 February 1942, HMS *Kedah*, being a local ship, was given the task of leading the convoy out through the minefields and the southern islands into the Durian Strait, then on to the temporary safety of Tanjung Priok in Java. Three and a half years later it was most fitting that on 5 September 1945 one of the first of Admiral Mountbatten's ships to enter Singapore's Keppel Harbour was HMS *Kedah*

In 1928 another Scot, Stuart Anderson, joined the firm and remained until his death in 1966, at which time he was senior partner. Fred Ritchie and Stuart Anderson escaped from Singapore just ahead of the invading Japanese forces in February 1942 on board Straits Steamship's *Vyner Brooke*. Unfortunately, the ship was bombed and sunk by Japanese aircraft in the Banka Strait and they spent the next three and a half years in a prisoner of war camp at Palembang in Sumatera.

While in the camp Ritchie frequently walked round the perimeter in the evening with a little Australian soldier. One day, in the course of conversation, Ritchie asked his Australian friend what line of business he had been in before the war.

'I was in the rice business,' was the reply.

'Oh,' said Ritchie. 'What part of the rice business were you in? Did you import it, or distribute it, or what.'

'What are you talking about?' said the surprised Aussie. 'I was a jockey for Christ's sake – I was in the bleeding *horse ricing* business!'

To keep themselves occupied in the camp, Ritchie and Anderson worked on plans for a new ship for Straits to replace the lost *Vyner Brook*. Their new design placed the bridge forward with the accommodation and engines aft, the bulky midship section being reserved for cargo. This would allow both hatches to be worked at the same time at the small berths in Borneo and avoid the tiresome business of warping the ship up and down the wharf during cargo operations. On their release from the prison camp in 1945 the plans were seen in London by the Straits Steamship Chairman and accepted. The ship was subsequently built by the Caledon Shipbuilding Company, almost exactly to the plans drawn up by the two marine consultants and carefully hidden from the Japanese guards in the Sumateran prison camp. She was launched in 1947 at Dundee and named *Rajah Brooke*. For the next 27 years she operated on Straits Steamship's Singapore-Borneo service, a magnificent tribute to Fred Ritchie and Stuart Anderson.

When I joined Ritchie & Bisset in November 1961 Stuart Anderson was the senior partner, Fred Ritchie having died in 1952. On my first morning in the office I was told that it would take me a couple of years to learn the basics of the surveying business. 'It's a whole new ball game, almost like starting your engineering apprenticeship all over again,' I was told. I didn't really believe them, but it turned out to be a fairly accurate forecast of my immediate future.

The work that I enjoyed most, after I had settled into this new life, was the investigation of marine casualties on behalf of the Salvage Association of London. This was an organisation founded in 1856 by Lloyd's and the major London insurance companies. Its purpose was to investigate machinery, hull and cargo damages and present underwriters with sound technical and financial opinions upon which they could judge the merits, or de-merits, of claims being submitted by shipowners. During the 1960s the Salvage Association did not have any offices in the Far East, although they had a resident representative in Japan, and all their survey work in Singapore and the adjacent territories was carried out by Ritchie & Bisset. Initially, most of the work I was involved in concerned run-of-the-mill engine and boiler damage, grounding damage and fire damage, with the repairs being generally carried out in the dry docks and workshops of the Singapore Harbour Board, now Keppel Shipyard.

Many of the ships I attended were war-built Liberty ships. In the early and mid 60s a casual glance around the

Singapore anchorage, or along the wharves of Keppel Harbour, would invariably reveal the distinctive silhouettes of numerous Liberty ships. They were by that time beginning to show their age, so it was not surprising that they were often involved in marine insurance claims of one sort or another.

The American Liberty ship was built to a British design for a simple cargo steamer capable of being produced quickly and in large numbers. In 1939 the Sunderland shipbuilding firm, Joseph L. Thompson & Sons, built *Dorington Court* for the Court Line of London. She was a single-screw steamer of economical design, a cargo capacity of 10,200 tons and a service speed of 11 knots. Main propulsion was by a triple-expansion, reciprocating, steam engine, developing 2,500 ihp at 75 rpm. Steam, at a pressure of 200psi, was supplied by two coal-fired Scotch boilers. This design was adopted by the Admiralty for their emergency wartime cargo-ship programme. One of the first ships to be built as part of this programme was named *Empire Liberty*. Detailed plans of this vessel were sent to Canada and the United States, where the British Government had placed an initial order for 60 ships. By the end of the war more than 2,500 ships had been completed in the amazingly short average working time of just 40 days per ship. In the Canadian West Coast shipyards they were built more or less as per the original design. In the United States they altered the accommodation layout by incorporating the bridge and all the accommodation in one midship house, located around the engine casing. They also fitted Babcock & Wilcox (B&W) oil-fired water-tube boilers instead of the coal-fired Scotch boilers of the original design. The reason for changing the boilers was the ready availability of the Babcock boiler in North America and the familiarity of American engineers with that type.

While working with Ritchie & Bisset at Singapore in the

1960s, I held appointments as a Non-Exclusive Surveyor to both the American Bureau of Shipping and Bureau Veritas. In this role I carried out countless Liberty ship boiler, machinery, hull, and safety-equipment surveys on their behalf, as well as insurance related surveys for the Salvage Association. Unfortunately, the records of those days – survey reports etc. – no longer exist. Indeed it is more than 40 years since I carried out my first survey on a Liberty ship; nevertheless some of the problems encountered remain fairly clearly etched in my memory.

There were, as I recall, two main problems with the B & W boilers in the Liberty. One was that repeated repairs to the plastic refractory on the furnace floor resulted in the floor level rising. This reduced the furnace volume and altered combustion conditions in the furnace, causing over-heating of the adjacent boiler tubes; these fire-row or screen tubes, as they were called, were frequently found sagged, or blistered, or both, at the time of survey. The Liberty used a degree of superheat in the HP cylinder of the main-engine and this meant that cylinder lubrication was required. Fail-ure to accurately monitor the rate of lubrication, and to check the feed water filters regularly for oil contamination, invariably resulted in an oil film being deposited on the waterside of the boiler, again causing overheating and dis-tortion of the fire-row tubes. On attending a boiler survey one first checked to see the condition of the feed filters. Then, on entering the furnace, a torch-light was shone along the underside of the fire-row tubes; this would clearly show up any distortion. The vertical distance from the centre line of the oil fuel burner/air-register assembly to the furnace-floor refractory was also checked. If the filters were clean, the fire-row tubes straight, and the height of the furnace floor correct, then you would probably not find much wrong with the boiler.

Another problem was waving or buckling of the bottom

plating at the forward end of the engine-room, in way of the boiler feed water double-bottom tanks. Nearly all the ships had a 'Condition of Class' in this regard. In other words, this area had to be specially examined at regular intervals in case of a possible deterioration in the condition of the bottom plating.

The two main problems that affected these ships in the immediate post-war years were the development of cracks on the foredeck and the loss of propellers resulting from fractured propeller shafts. Both of these serious problems had been dealt with before I became involved, but at class surveys particular attention was always paid to the deck around the hatch corners, especially No. 3 hatch, and at the bridge front. Even with reinforcing straps fitted, Liberty ships could develop serious fractures in that area if they encountered exceptional heavy weather conditions.

Ritchie & Bisset carried out numerous On and Off-Hire Surveys on behalf of various interests. I was always happy if the vessel was a Liberty as frequent visits to this class of ship meant that you could remember the bulkhead and some of the frame numbers without having to refer to the ship's drawings. This saved quite a bit of time and trouble. The American Bureau of Shipping had a mini General Arrangement plan for a Liberty, which you could keep in your notebook and refer to should your memory fail.

At this stage operating a Liberty ship was beginning to prove uneconomical, as it was becoming more and more expensive to maintain them in accordance with class requirements. They were also uneconomical as regards fuel consumption when compared with a modern diesel-powered vessel of similar cargo-carrying capacity. A 'Total Loss', or 'Constructive Total Loss', situation was probably the most financially attractive method of disposing of a Liberty ship at this time. The term 'Total Loss' is self-explanatory – the ship has been destroyed. A 'Constructive Total Loss' is one where the ship, while not being destroyed, is nevertheless so badly damaged, or is so inaccessible from a salvage point of view, that the salvage and repair costs will exceed the insured value. Some of the more marginal operators may well have been tempted to deliberately arrange an insurance fraud along these lines in order that they could get rid of their uneconomical Liberty ship and pocket the insured value, or invest in a more modern vessel. Not surprisingly underwriters very often viewed this type of casualty with a somewhat jaundiced eye.

However, a deliberate 'Total Loss' is not an easy operation to arrange. There is a very real risk to life, whether it be a grounding, a scuttling, or an engine-room fire. In addition, some of the crew will know, or suspect, what is going on and may give the game away. Because of these factors this sort of insurance scam was not very common. What was much more common was that where a Liberty had sustained reasonably extensive genuine damage, the owners would try very hard to prove to the underwriters, by fair means or foul, that the cost of the damage repairs was going to exceed the insured value. By this method they would hope to obtain a settlement with the underwriters on the basis of a 'Constructive Total Loss'. Needless to say, this involved the owners' and the underwriters' surveyors in some fairly acrimonious discussions.

During 1967 and 1968 I was involved in one 'Constructive Total Loss' case, and two 'Total Loss' casualties on behalf of the underwriters. Despite the absence of documentary records, and the passage of more than thirty years, they have remained fairly clearly imprinted in my memory. Nevertheless I have had to check names, dates and times with old Lloyd's Shipping Index and Lloyd's List Casualty records at the Guildhall Library in London.

The first concerned the Liberian flag *Angelina*, which started life as the Ministry of War Transport's *Samsacola*, built by Bethlehem Fairfield Shipyard Inc. at Baltimore in 1944.This vessel was originally managed for the Ministry by Silver Line of London. In 1947 she was sold to the managers and renamed *Silvercedar*. In 1949 Silver Line sold the vessel to Ben Line and she became *Benwyvis*. Ben Line operated her on their Far Eastern Service until 1955 when she was sold to Liberian/Greek interests. She operated as *Linda* until 1958 and then as *Agia Irene*. In 1965 she became *Angelina* with registered owners as Transocean Navigation Corporation of Monrovia.

On 31 July 1967, *Angelina* was on a ballast passage from Vishakhapatnam to Madras to load iron ore for Japan, when a serious fire occurred in the engine-room.

The casualty was initially reported to Lloyd's by their agent at Madras as follows:-

Madras 01 August

Steamer Angelina *reported abandoned. Master and 24 crew rescued by British steamer* Golden Phoenix. *Understand one fireman missing.* Golden Phoenix *due to arrive at Madras on 01 August.*

Lloyds Agent.

Madras 02 August

Steamer Angelina *en route Vishakhapatnam to Madras in ballast to load iron ore for Japan reported on fire at 2.30pm on 31 July at latitude 14 degrees north, longitude 81 degrees 26 minutes east. Approximately 85 miles ENE of Madras. Indian steamer* State of Assam *proceeded from Madras at 8.00pm 31 July to investigate and report. Meanwhile Port tugs held in readiness to proceed if required.*

Lloyds Agent

Madras 04 August

Angelina *reported afloat but still burning.*

Lloyds Agent

Madras 08 August

Tug Friesland *reported at 8.30pm 04 August –* Angelina *afloat in latitude 15 degrees 28 minutes north, longitude 85 degrees 19 minutes east. Approximately 300 miles ENE Madras. Tug expects complete tow arrangements on 06 August and will bring* Angelina *to Madras.*

Lloyds Agent

The master of the salvage tug *Friesland* initially intended to tow the casualty to Madras outer anchorage for inspection and redelivery to owners. However, Madras Port Trust were not willing to accept the casualty as the imminent onset of the monsoon would make it unsafe for a disabled vessel to remain at the anchorage, and severe port congestion meant that they were unable to provide a suitable safe berth. In view of this situation the tug, after consulting with owners, decided to tow the casualty to Singapore, where there were no problems with either a safe anchorage or a suitable lay-by berth. *Friesland* with *Angelina* in tow arrived safely at Singapore on 17 August and went to anchor at the Eastern roads.

At Ritchie & Bisset we received instructions from the Salvage Association, on 16 August, to attend the casualty on behalf of the underwriters concerned after arrival at Singapore, and ascertain the cause, nature and extent of the damage, together with the estimated cost of the necessary repairs. In this connection I boarded the vessel at the Eastern Anchorage on 20 August, in the company of a marine consultant who had flown out from New York to look after the owner's interests.

Our inspection showed that the entire midship house was completely gutted, with all decks, bulkheads and casings heavily distorted. In the bridge area all the navigational and radio equipment had been destroyed. The port and starboard side shell plating of the engine-room was heat-buckled in way of the upper three strakes. The hatch-boards at No. 3 and 4 holds were completely destroyed together with all the wood sheathing in the lower part of the two holds. The 'tween deck plating in way of both holds was generally heat-buckled together with the associated underdeck beams and brackets.

Our first day was spent carefully noting down the exact extent of the hull and superstructure damage. On the second day a detailed inspection of the engine-room was carried out to establish the extent of damage sustained by the machinery and, perhaps more importantly, to endeavour to ascertain the source of the fire.

Our inspection of the decks had shown deposits of partly burned fuel oil around the starboard fuel settling-tank airpipe. This suggested to us that the tank had overflowed at some time prior to the fire. On entering the engine-room we therefore directed our attention, in the first instance, to the starboard settling-tank. Heavy deposits of carbonised and partly burned fuel oil could be seen both around the tank manhole cover and right down the inboard side of the tank. Similar deposits were also found on the adjacent rear casing of the starboard boiler. In addition, the tank manhole cover securing nuts were found to be loose.

On the Liberty ship the fuel oil settling-tanks were wing tanks located on the port and starboard sides at the forward end of the combined engine and boiler room. The Babcock & Wilcox boilers faced inwards towards each other, with the fuel oil pressure pumps and heaters located between the boilers. The rear of each boiler was adjacent to the inboard side of the settling-tank, but separated from it by a two-feet wide walkway. Any heavy fuel leakage spilling down the inboard side of a settling-tank would almost certainly come into contact with the rear casing of the adjacent boiler.

Our findings fairly clearly indicated that the starboard settling-tank had been over-filled, causing fuel oil to spill out of the unsecured manhole cover on the top of the tank, flow down the inboard side of the tank, and splash onto the adjacent rear casing of the starboard boiler. Any hot-spots on the boiler-casing as a result of missing or defective refractory brickwork would have caused auto-ignition of the spilled fuel oil. The casings of the Babcock & Wilcox boiler were lined on the inside with fire refractory material, which prevented radiation and kept the outside of the casing quite cool. However, despite annual boiler surveys by classification surveyors, the effects of more than twenty-odd years of wear and tear sometimes resulted in local failure of the refractory lining, leading to hot-spots on the casing – hot enough in certain circumstances to ignite spilled fuel oil.

The next two or three days were spent reaching agreement on the extent of the repairs necessary to return the vessel to its pre-casualty condition. Once that had been done we then looked at the probable cost of the necessary repairs. This exercise quickly showed that our agreed estimated damage-repair cost would exceed the vessel's insured value.

At the end of our inspections and deliberations the underwriters were advised that the fire had most probably been caused by a combination of accidental overfilling of the starboard settling-tank, and crew negligence in failing to ensure that the settling-tank manhole cover was correctly secured. They were further advised that it would not be possible to repair the fire damage within the insured value.

The result of our investigation was telexed to London

and can be read, except for the information concerning the estimated cost of repairs, in Lloyd's List of 1 September 1967.

Singapore 31 August

Steamer Angelina *Surveyor reports that investigation reveals fire started after overfilling starboard fuel oil settling-tank which then overflowed on main deck. Fuel also overflowed down inboard side of settling-tank via the manhole cover which found improperly secured. Fuel most probably ignited by the leakage from the manhole cover coming into contact with overheated section of the starboard boiler rear casing.*

Lloyds Agent per Salvage Association.

Several weeks later *Angelina* was towed off to the ship breakers for demolition, the underwriters having apparently settled with the owners on the basis of a 'Constructive Total Loss'.

The first Liberty ship 'Total Loss' that I was involved with concerned the Greek flag vessel *Kostis A. Georgilis* which I inspected at Great Coco Island. This island lies just to the north of the Andaman Islands, and some 250 nautical miles south-west of Rangoon. It is part of Myanmar, or as it was then called, the Union of Burma. The ocean route from Rangoon to Dondra Head (southern tip of Ceylon) passes just to the north of the island.

Kostis A. Georgilis was built for the United States War Shipping Administration by Bethlehem Fairfield Shipyard Inc. at Baltimore in 1944 and delivered on bare boat charter to the Ministry of War Transport as *Samconstant*. In 1947 the vessel was sold to commercial interests and served several owners, first as *Skipsea* until 1948 and then *Ramon de Larrinaga*. In 1952 the vessel was bought by Okeanoporos Shipping Corporation of Panama and operated as *Okeanoporos* until 1962 when she changed name to *Kostis A. Georgilis*. Although the registered owners were in Panama, the vessel operated under the Greek flag, and the real owner was probably the master, Captain Kostis A. Georgilis.

The ship arrived in the Rangoon River, on completion of a ballast passage from Chinwangtao, on 4 October 1967. After waiting at the inner anchorage for several days, the vessel moved up river to a double-mooring in the stream off Sule Pagoda Wharf, where a full cargo of bagged rice was loaded from barges. Cargo loading was completed on the morning of 2 November and the vessel sailed for Colombo later on the same day.

Approximately 24 hours after dropping the pilot at the mouth of the Rangoon River, a fire occurred in the engine-room close to the boiler fronts. The fire spread rapidly and the engine-room staff were forced to abandon the machinery space. At this time the vessel was close to Great Coco Island and the master decided to alter course towards the island with the intention of beaching the vessel should the fire get completely out of control. Although the engine-room had been abandoned and was ablaze, the main-engine and steering gear were still functioning. As the ship closed on the island the fire spread into the accommodation; this caused the master to decide to run his vessel ashore while he still had main-engine power available.

After beaching the burning ship, the crew, which included the master's wife and daughter, all got clear in the ship's lifeboats and landed safely on the shore. They had taken the emergency radio transmitter with them and once ashore set it up ready to transmit a distress call. Before abandoning the vessel a 'May Day' call had been transmitted by the

47

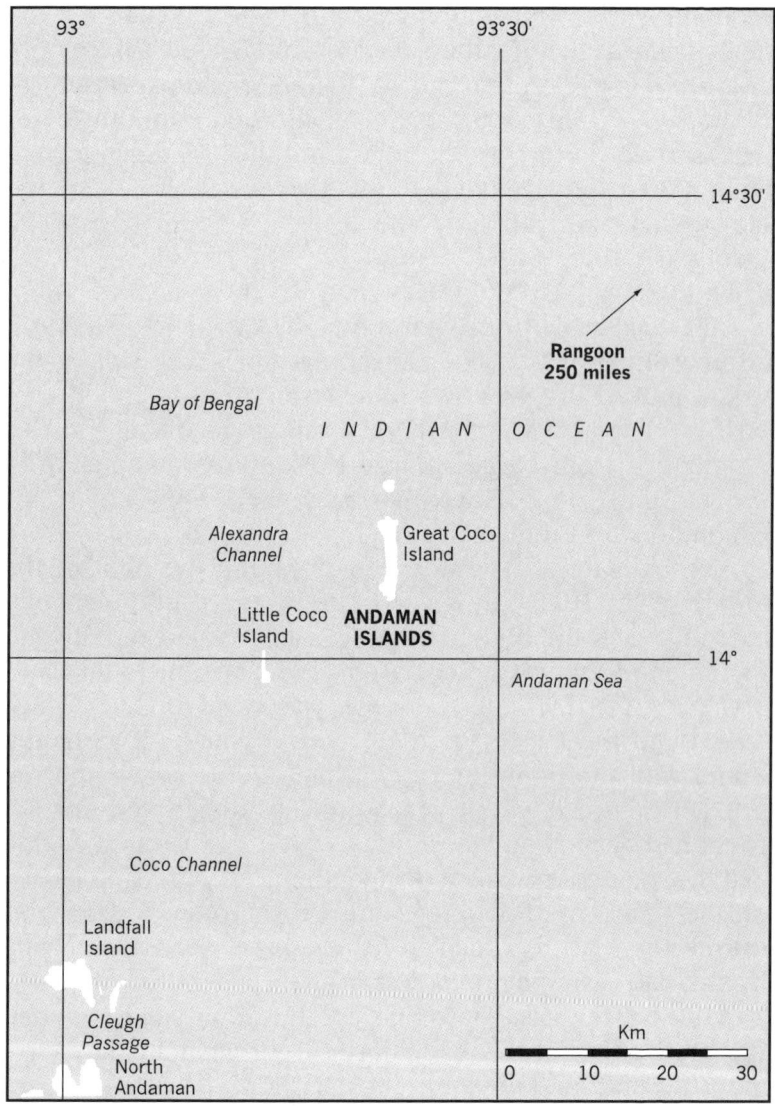

Map 1 Great Coco Island and the northern end of the Andaman Islands.
The Myanmar/Indian border lies just north of the Coco Channel at
13 degrees 50 north

48

radio officer but had not been acknowledged by any ship or shore station, and they were therefore anxious to send out another. However, before they could start to transmit, a Burmese Navy detachment descended on them, confiscated the radio and placed them all in custody. Unfortunately for them, Great Coco Island was a very restricted area, containing a detention camp for political prisoners, and was strictly 'out of bounds' to all but authorised Burmese military personnel.

All through the night of 3-4 November the beached ship burned fiercely. Debris from the burning accommodation house set the No. 3 and 4 hold hatch-covers and hatch-boards alight and the fire then spread to the rice cargo as burning pieces of hatch-board dropped into the holds onto the rice bags. The fire later spread to No. 2 hold. For several hours after the beaching there was still sufficient steam in the boilers to turn over the engine and, at low water, the crew watched the propeller slowly rotate, throwing up spray around the stern, until, eventually, the steam pressure died away and the propeller stopped.

The master, his family, and the crew spent two rather miserable weeks in a couple of wooden huts that the Navy made available for them. Food, mainly rice, was also provided. Washing and toilet facilities were basic in the extreme and at night mosquitoes kept them all awake. The Navy had, of course, advised their headquarters in Rangoon of their unexpected guests and this information filtered through to both the charterers, who were the Burma Rice Board, and the owner's Rangoon agent. Soon the owners in Piraeus were talking to their insurance brokers and the underwriters in London, and as a result the underwriters appointed The Salvage Association to investigate the situation on their behalf, they in turn appointed Ritchie & Bisset to carry out the survey and investigation.

On 7 November I was instructed by the Ritchie & Bisset

senior partner to proceed to Rangoon and, with the assistance of the Lloyd's agent, arrange onward transportation to the casualty at Great Coco Island. My instructions were to ascertain the cause, nature and extent of the damage, the prospects of refloating and towage to a port of repair, and the estimated cost of salvage and repair. Travel to Burma required a visa and in, those days, the completed visa application form had to be sent from the Burmese Embassy in Singapore to Rangoon for approval before the visa could be issued. This procedure took almost a week and it was 14 November before the visa was stamped in my passport.

I arrived in Rangoon on the morning of 15 November where I was met by Mr U Shwe Yee, of Golden Bird Agencies, the local Lloyd's agent, who had arranged for me to meet with representatives of the owners, the charterers, and the Burmese Navy. After the meeting, the Burmese Authorities gave us clearance to fly down to the casualty location the following morning on a Union of Burma Airline DC3. This was a very large aircraft for just four passengers, but it was a 500 mile round-trip over water all the way, so the big DC3 made us all feel nice and safe.

Great Coco Island is almost 6 miles in length in a north-south direction, and just under a mile wide. The western side of the island rises fairly steeply from the sea, but the eastern shore is virtually flat with just a gradual slope to the sea. There are no off-lying dangers, apart from rocky and foul ground extending some three cables from the north and north-western shore. Except where it has been cleared to build the naval camp, the detention camp and the airstrip, the island is covered with dense rainforest.

The flight took just over an hour and as we circled the island prior to landing we passed over the beached ship and could see plumes of smoke rising from the area of the No. 3 and 4 holds. It was mid-morning when we disembarked and introduced ourselves to the naval officer in charge. The

Navy, who had been informed of our arrival, had arranged for us to be transported to the beach opposite the casualty. This was at Ford Bay on the eastern side of the island and about two miles south of the northern tip. The transport consisted of an agricultural type of trailer, on which we sat, rather like visitors to an old-fashioned Country Fair, towed by a tractor. At the beach we met the master, chief engineer, and the mate, who, I need hardly say, were extremely glad to see us. By that stage they had had more than enough of Great Coco Island. They had their motor lifeboat ready and, after I explained what I wanted to do, they took me out to the casualty.

Kostis A. Georgilis was aground in geographical position 14 degrees 07.5 minutes north, 95 degrees 22.5 minutes east. She lay about 200 metres off shore, down by the stern and with a starboard list.

The stern was in approximately 25 feet of water and at high water there was some movement in the after half of the vessel. The fore end was firmly aground and almost dried out at low tide. As we approached we could see that the midship house was completely burned out. Large areas of paint-work on the side shell-plating in way of the engine-room and Nos. 2, 3 and 4 holds had been burned off, with the steel-work in way heat-buckled to varying degrees. The acrid stench from the burned paint, accommodation and rice cargo hung over the ship and the surrounding area. It was a smell that I was to become quite familiar with over the next 30 years.

The starboard lifeboat falls were hanging down into the water and we were able to scramble onboard fairly easily. The interior of the midship accommodation house was completely destroyed. Moving along the working alleyway I came to the engine-room door and stepped inside onto the upper gratings. Above me the engine-room casing and skylight was heat-deformed and smoke-blackened, while

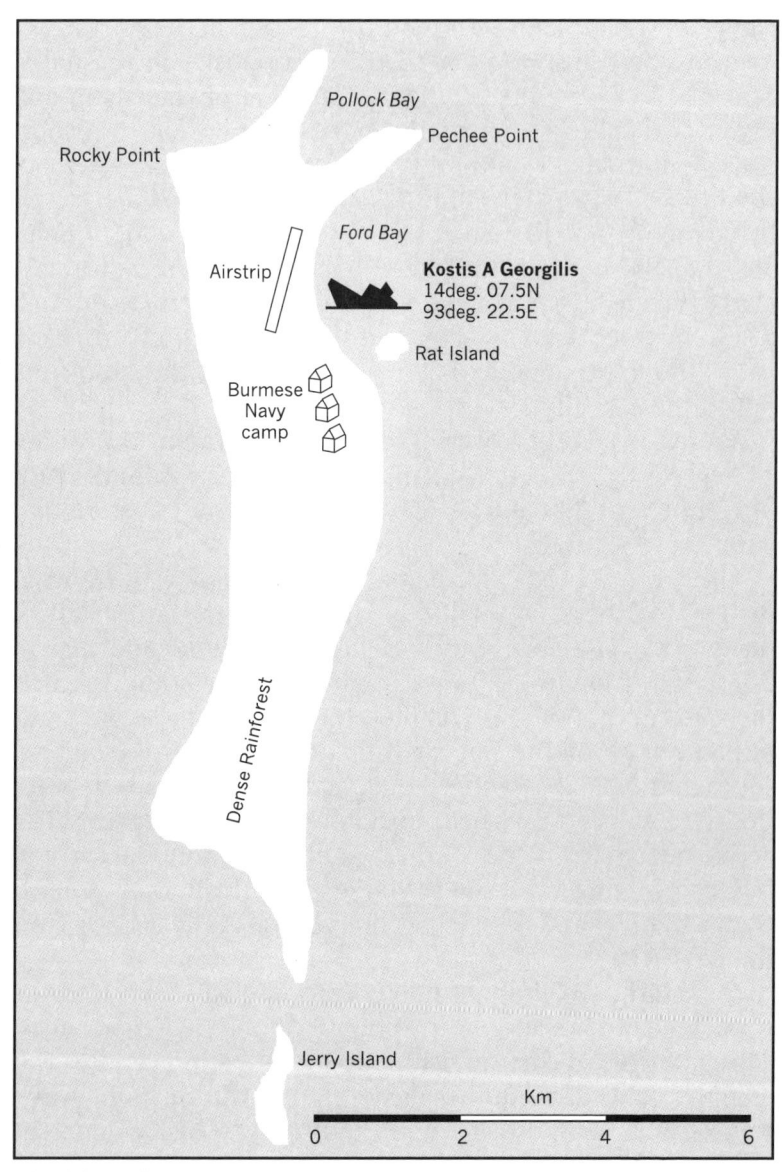

Map 2 Great Coco Island

about five or six metres below me an inky liquid surface sloshed gently from side to side: the engine-room was partly flooded with fuel oil. Under these conditions there would be no chance of inspecting the boilers and, in particular, the fuel oil burning system, in order to establish the cause of the fire. In the working alleyways the main-deck plating was heavily buckled. In the forward cross alleyway the remote closing devices for the fuel oil burning system, and the controls for the engine-room steam-smothering system, were heavily distorted. Out on deck the weather-deck plating in way of Nos. 2, 3 and 4 holds was distorted by the heat from the burning rice and was, in fact, quite hot to the touch. On approaching the No. 3 weather deck hatch I was met by a wave of heat like the blast from a furnace. Looking down into the hold, all I could see was an incandescent glow from the burning rice. A similar situation existed at the Nos. 2 and 4 holds. In all I suppose that I spent about two or three hours looking round the casualty before returning ashore in the ship's lifeboat.

There was no doubt that *Kostis A. Georgilis* was well beyond economical salvage and repair. The vessel had been reduced to a burned-out hulk. I cannot now recall what estimated cost I subsequently passed to the underwriters but it would have been well in excess of the insured value.

Once ashore I advised the Burmese Navy of my findings and prepared to return to Rangoon. At this stage the master asked if it would be in order for the crew, his family, and himself to return on the aircraft with us. I assured him that there was absolutely nothing that he could achieve by remaining on the island. By this time the Burmese authorities, influenced perhaps by the presence of the captain's wife and little daughter, had decided the crew were harmless, and we were all able to fly back to Rangoon together in the DC3. The Burmese Navy were quite happy, after all they had aquired a motor lifeboat and an emergency radio.

No doubt they thought they would prove to be useful pieces of kit for them.

As we flew out of the island we again passed over the vessel. Columns of smoke were still rising from the burning rice and I could not help thinking what a sad end to the old ship. Flying back to Rangoon I wondered about her early career when she had flown the Red Ensign as the *Samconstant*. Had she been in the North Atlantic convoys, or the run to Murmansk, or been part of the D-Day armada in the Channel on 6 June 1944? Perhaps in these very waters she had carried supplies for the 14th Army at the recapture of Akyab and Rangoon in 1945. What history had the old ship helped to make, I wondered.

Once back in Rangoon I sent off a telegram to London advising the underwriters of my findings. In those days there were no telex facilities available in Burma and telegrams had to be handed in personally at the Rangoon General Post Office where they were vetted to see that they contained nothing that was likely to prejudice the security of the State. I see from the Lloyd's List records in the Guildhall Library that my first advice to the underwriters read as follows:-

Rangoon 20 November

Boarded vessel today and found cargo in holds still burning making parts of vessel inaccessible. Main deck very heavily buckled over full length and 'tween deck likewise. Port and starboard shell plating generally distorted over full length from main deck to light load line. Midship house completely gutted together with upper engine room and boiler room. Lower engine room flooded with fuel oil. Cause of fire as yet unknown and still under investigation.

Salvage Association Special Surveyor.

The underwriters had advised that they were suspicious of the circumstances surrounding the casualty, and I was requested to pay particular attention to this aspect. As already mentioned, with the engine-room partly flooded with fuel oil there would be no opportunity of examining the area round the boilers to try and ascertain how the fire had started. All I could do was to interview the crew and establish the exact sequence of events surrounding the outbreak of the fire. Once that was done it might be possible to throw some light on whether or not the loss of the vessel was totally accidental in nature.

On the morning after our return from Great Coco Island I started to interview the crew in my room in the Strand Hotel. Fortified by jugs of coffee carried up from the kitchen at frequent intervals by silent-footed Burmese waiters, we toiled away for the best part of two days trying to piece together the events of the short fateful passage from Rangoon.

The information that emerged was, that after completing cargo loading the vessel departed on the morning of 2 November, and was 'Full Away On Passage' for Colombo at about noon. All operations of the vessel appear to have been normal for the remainder of the day and through the night of 2-3 November, in fact, right up until the time of the fire, which, as far as I could establish, was close to 11.00 a.m. At that stage the vessel was almost 24 hours into the passage to Colombo and was approximately 15 miles north east of Great Coco Island, the master's intention being to follow the normal ocean route between Rangoon and the southern tip of Ceylon.

I was particularly interested in the evidence of the watch-keeping engineer and fireman. What were they up to before the fire started? The answer to that question was: nothing unusual really. Shortly after the start of the 8-12 watch they changed over the filters on the boiler fuel system, cleaned the filter elements on the filters that they had taken out of service and then replaced them. They also changed over the fuel supply to the boilers from the starboard to the port settling-tank. Some time later, the fireman started up the fuel transfer pump in order to fill the almost empty starboard settling-tank from one of the double-bottom fuel tanks. This was at around 9.00a.m., according to the fireman. As already described the fuel oil settling-tanks on the Liberty ships were wing tanks located on the port and starboard sides at the forward end of the combined engine and boiler room, adjacent to the rear casings of the boilers. The Babcock boilers faced inwards towards each other with the fuel oil pressure pumps, heaters, and filters located between the boilers. One set of filters, the final hot filters before the fuel burners, was located adjacent to the boiler front.

Shortly after 11.00a.m. the engineer and fireman were both on the main-engine mid-level grating 'feeling round' the crosshead bearings and attending to their lubrication. The method of checking the temperature of the bearings on the triple-expansion steam engine of the Liberty ship was simply to touch them by hand. This required a fair degree of co-ordination between eye and hand if you wanted to avoid losing a finger or two. You could, of course, spit on them; if they spat back they were too hot! Anyway, while inspecting and/or oiling the HP cylinder and valve gear bearings, they suddenly became aware of flames dancing about the furnace front of the starboard boiler and around the fuel oil heaters. Now what happened next is not very clear. After advising the bridge and chief engineer of the situation, an attempt was made to attack the fire using the ten gallon, semi-portable, foam extinguisher located on the lower grating between the engine and the boilers. This, for whatever reason, was not successful; the crew said that the fire, which had spread rapidly and covered the entire front of the starboard boiler and the fuel oil pressure pump and heaters, had threatened to engulf them as they tried to direct the foam jet towards the fire. At this stage it would appear that, with the fire spreading rapidly, the engine-room was abandoned, with the boilers steaming, and the main-engine, generator and steering gear still running.

The action that the engine-room crew should have taken, once it became clear that they were unable to control the fire locally, was to shut down the main-engine and close the fuel supply to the boilers by closing the starboard settling-tank outlet valve. The steam supply valves to the fuel oil burning system, and the forced draught fan should also have been closed. These valves were provided with extended spindles so that they could be operated from outside the engine-room in just such an emergency. With the fuel supply isolated, the engine-room vents and skylights should then have been closed and the steam-smothering system activated so as to cut off the air-supply to the fire. This is done by opening an isolating valve on the boiler drum, thus allowing steam to flow through perforated pipes which run around the bottom of the boilers and the fuel oil burning equipment. After only a few minutes the lower half of the boiler-room should have filled with steam, thereby suffocating the fire. This valve could also be operated from outside the engine-room by means of an extended spindle. The extended spindle operating hand-wheels for both the fuel and steam isolating valves and the steam-smothering controls were, as has already been mentioned, all located at main deck level in the forward cross alleyway.

The fuel supply was not shut off, apparently because the master wished to steam towards the island and therefore wanted to retain the use of the main-engine for as long as was possible. The engineers said that they opened the steam smothering valves on both boilers but this would appear to have had no effect on the fire. This did not surprise me too much. My experience of carrying out safety-equipment surveys on ships of this type was, in many cases, that the remote controls for the fuel oil cut-off and steam-smothering valves could not be operated from the remote location in the main deck cross alleyway. This was usually because the extended spindles were either disconnected or seized. In fact, they were usually disconnected because they were seized. This was especially the case with the fuel supply valves. If the extended spindles of these valves were connected, but covered in paint, grime and rust, instead of being clean and properly lubricated, they were very often difficult to operate locally in the engine-room when the engineers were changing over settling tanks. The immediate answer was simple: the extended spindles were discon-nected, and nobody would remember to overhaul and reconnect them until the next safety-equipment survey.

While on board the casualty, I had inspected the remote fuel and steam-smothering valve controls in the cross alley-way. None could be moved, but the extended spindles would all have been badly damaged in the fire so the fact that they would not move at that stage told me nothing about their condition prior to the fire. Owing to the damage in the cross alleyway – in particular the distorted condition of the extended spindles, hand-wheels and adjacent deck plating, I was unable to ascertain whether or not the valves were open or closed.

One important point that arose during the interview with the 8-12 engine-room watch keepers was the transfer of fuel into the starboard settling-tank. Was this operation com-pleted before the fire was detected? If it was started at around 9.00a.m., as the crew stated, it should have been finished by 10.30a.m. at the latest. The engineer stated that he shut down the transfer pump when the tank was full shortly after 10.00a.m., he thought. If, in fact, the fuel transfer pump had been forgotten about then the starboard settling-tank would have overflowed some time between

10.30a.m. and 11.00a.m., which was just about when the fire started. The two witnesses to the start of the fire were, however, quite adamant that the first flames had been in the area of the fuel oil heaters and furnace fronts of the starboard boiler, *not* the rear of the boiler, which would have been the case had it been associated with an overflow situation at the starboard settling tank. At the time that the fire was first detected the engineer and fireman stated that they were on the main-engine mid-level grating beside the HP cylinder and valve gear crosshead bearings. This is at the forward end of the main-engine, immediately behind and above the boiler fronts and the fuel oil burning equipment. They would, therefore, have had an excellent view of the area between the boilers and the furnace fronts. If their evidence is to be believed, then it would appear that the fire was associated with fuel leakage on some part of the fuel burning system close to the starboard boiler furnace fronts, possibly the hot fuel filters.

At the end of the day I did not obtain any real evidence as to the prime cause of the fire, or whether or not the loss was deliberate. The actual cause could only be accurately determined by a careful examination of the fuel transfer pump, the starboard settling tank, and the starboard boiler furnace fronts – in particular the high-pressure hot fuel filters. However, with the lower engine-room flooded this would not be possible. The fact that the master had his wife and small daughter on board made it, in my opinion, unlikely that he would engage in a deliberate 'Total Loss' claim – especially one that involved a potentially dangerous fire and stranding at a remote island.

Perusal of the Lloyd's List records at the Guildhall Library showed that at the time I must have accepted the evidence of the engine-room watch keepers regarding the start of the fire and the filling of the settling-tank, as my second message to the underwriters read as follows:-

Rangoon 21 November

Steamer Kostis A. Georglis. *Fire reported to have originated at forward side starboard boiler adjacent to hot fuel
oil filters and probably due to leakage from fuel pressure
pipe or filters coming into contact with hot boiler casing.
Fire spread rapidly causing engine room to be abandoned
within 10 minutes with main engine and auxiliary units
still running. Vessel presently 200 yards from shore with
both anchors down and bow firmly aground but aft half
afloat at high tide. As yet no signs of leakage. Partial
flooding of engine room due early extinguishing efforts
and internal leakage from fuel oil settling tanks. Present
indications are that fire in holds will continue for perhaps
another week. All crew including Master now at Rangoon.*

Salvage Association Special Surveyor.

In addition to the foregoing text the message would have
contained 'Confidential Advice to Underwriters' that the
cost of salvage and repair would greatly exceed the insured
value of the vessel. That sort of information was not published in Lloyd's List.

London solicitors acting for the underwriters were for,
some reason or other, convinced that this had been a
deliberate loss, and correspondence continued for many
months. Attempts were made to engage a salvage contrac-
tor to attend at the casualty location and pump the fuel oil
out of the engine-room in order that I could carry out a
thorough examination. However, the Burmese Authorities
did not want any foreign salvage craft entering their
restricted area around Great Coco Island and the proposed
inspection had to be abandoned.

Looking at this case now, after more than 30 years
experience of investigating engine-room fires, I find it
strange that fuel leakage occurred some three hours after

the filters were reported to have been changed over. I would have expected the leakage, usually the result of a defective filter-cover gasket or improperly secured cover, to have occurred when the filters were changed over, not some three hours later. Also, if the leakage did in fact occur when the filters were being changed over it is more than likely that the engineer, or fireman, doing the change-over would have sustained some sort of burn-related injury. Of course it is quite possible that a gasket on one of the filters could have blown out some time after the filters were changed over and when no one was standing close to it.

Perhaps the underwriters were justified in being uneasy about this case. Was it just a coincidence that the fire occurred conveniently in the middle of the morning when the ship was nicely positioned close to the island? It also seemed a little strange that, with most of the crew available at that convenient hour, there seemed to have been no serious attempt to fight the fire. From that point of view it did look a bit odd. But still, would the master have really deliberately placed his wife and little daughter in danger? I don't really think so. The balance of probabilities is, as I said, that the fire was the result of leakage from some part of the boiler fuel oil burning system – probably the hot filters on the starboard boiler.

The other Liberty ship 'Total Loss' that I was involved with was *Universal Trader*. This was a Liberian flag vessel with registered owners as Universal Shipping Co. Ltd., Monrovia. She was built for the United States War Shipping Administration by J.A. Jones Construction Co., Panama City, Florida, in 1944 and entered service as *Edward K. Collins*. In 1947 the United States Administration sold her to Greek owners and she was renamed *Chelatros*. A further change of name and owner occurred in 1961 when she became *Souliotis II*. Ownership passed to Universal Shipping Co. Ltd. in 1963.

Universal Trader arrived in ballast at Gdynia in Poland from Rostock on 29 December 1967 and, after loading a full cargo of grain, sailed for Chittagong, via Las Palmas and the Cape, on 14 January. She was reported at Lloyds as having sailed from Las Palmas on 26 January. The next report, dated 9 March, advised that she was aground and on fire in the engine-room on the Little Basses reef off the south-east coast of Ceylon.

On 11 March Lloyd's agent reported as follows:-

Colombo 11 March

Steamer Universal Trader – *Following received via Master* Ocean Enterprise – *Vessel stranded on rocky bottom with rough sea and swell. Cloudy with rain patches. Forepart DB tanks settling tanks and dry tanks full of water and fuel leaking out.*
No. 1 hold flooded to depth of 16 feet. No. 2 hold flooded to depth of 14 feet. No. 3 hold flooded to depth of 21 feet.
Visual wrinkles and cracks both sides of engine room plating. Water and oil leaked into engine room from settling tanks and caught fire first in engine room and then flames spread to accommodation. Dense smoke from hatches covered deck. Fire fighting impossible in such conditions and we abandoned vessel.
Vessel carrying grain – previously reported as coal.

Lloyd's Agent.

Following the stranding, the casualty broke in two at the engine-room forward bulkhead and the fore part flooded throughout and slid partly off the reef into deeper water. The underwriters, on being advised of a potentially serious claim, appointed Salvage Association to attend; they in turn appointed Ritchie & Bisset. Once again I found myself on the way to the airport, bound, on this occasion, for the beautiful Indian Ocean island of Ceylon. In 1972 the island reverted to its ancient name of Sri Lanka but at this time, 1968, Ceylon was still the official name.

I arrived at Colombo on 17 March with the usual instructions to investigate the cause, nature and extent of damage and the prospects of successful salvage. I was also requested to investigate any suspicious events associated with the casualty. On arrival a representative of Aitken Spence & Co., Lloyd's agents for Ceylon, met me and whisked me off to their office.

After discussing the situation it was agreed that one of their senior staff, Willie Vandenberg, would accompany me to the Little Basses round on the south-eastern coast, some 150 miles from Colombo, to act as guide, interpreter, etc.

The south-western end of the Little Basses ridge lies about one mile off Butawa point and some 30 miles east of Hambantoto. The ridge then runs in an east-north-easterly direction for 17 miles, more or less abeam the Yala National Park and Game Reserve. The ridge, which is very narrow and steep-to, has many rocky heads with depths of from 9 to 18 feet over them and on which the sea often breaks. The Little Basses reef lighthouse is located close to the north-eastern end of the ridge. A similar ridge, the Great Basses, lies further to the south-west.

To get to the casualty location we proceeded by car from Colombo down through the old port of Galle, past the lighthouse at Dondra Head on the southern tip of Ceylon, then on through Hambantota to the game reserve, where we spent the night in a very comfortable chalet.

Accommodation-wise it was to be downhill all the way after that first night.

For the next stage of the journey we had to enlist the help of what was then the Ceylon Lighthouse Service. Back in Colombo, Aitken Spence & Co. had arranged for one of the lighthouse boats to pick me up on the beach about a couple of miles from the game reserve chalet, and take me out to the casualty. First thing on the morning of 19 March one of the game reserve Land Rovers took me down to the beach where the boat, *Pharos II* was waiting. The drive through the game reserve was fascinating. There were leopards sunning themselves on rocks in the early morning sun, colourful peacocks perched on tree branches and, at one stage, we were brought to a halt when we met an elephant strolling up the track in the opposite direction. He stood contemplating us for about ten minutes, with his trunk waving gently from side to side, before losing interest and lumbering off into the bush.

Pharos II turned out to be a converted ship's motor-lifeboat in charge of a lieutenant of the Lighthouse Service and a crew of four seamen. Their normal duty was the ferrying of stores and personnel from their base at Hambantoto, some 35 miles to the east, out to the lighthouses on the Great and Little Basses. The trip out to the casualty took just over an hour; the weather was fine and clear, but there was a fairly heavy swell running. As we approached the stranded vessel we noticed several small local craft alongside with quite a number of people clambering about the decks and accommodation. Whoever they were they had us well and truly outnumbered.

'Bloody looters,' said the lieutenant. 'I'll soon get rid of them.' One of the seamen passed him a long canvas bag from which he withdrew a .303 Lee Enfield service rifle.

64

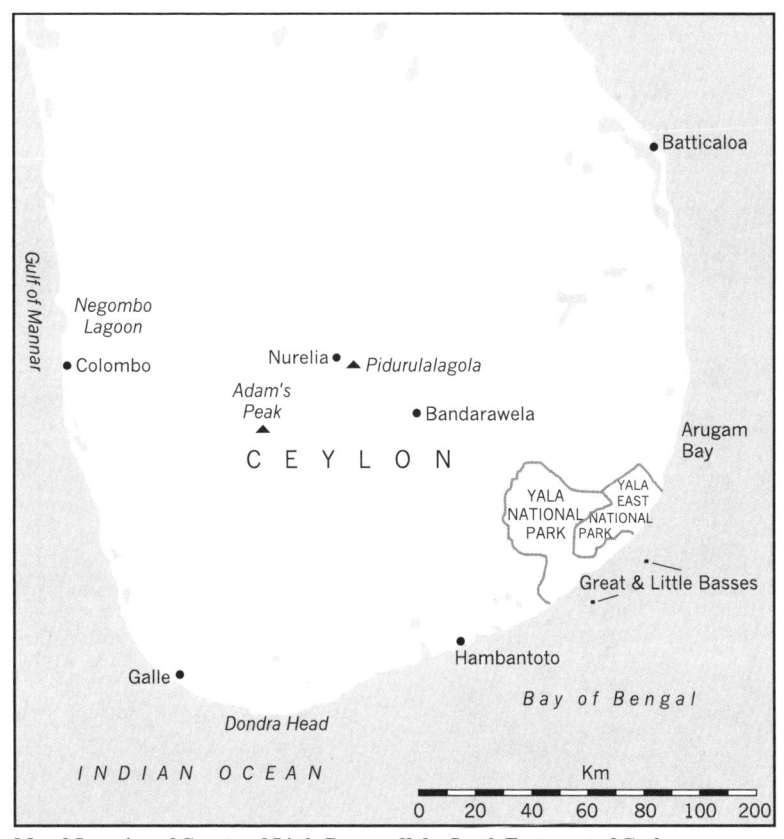

Map 3 Location of Great and Little Basses off the South East coast of Ceylon

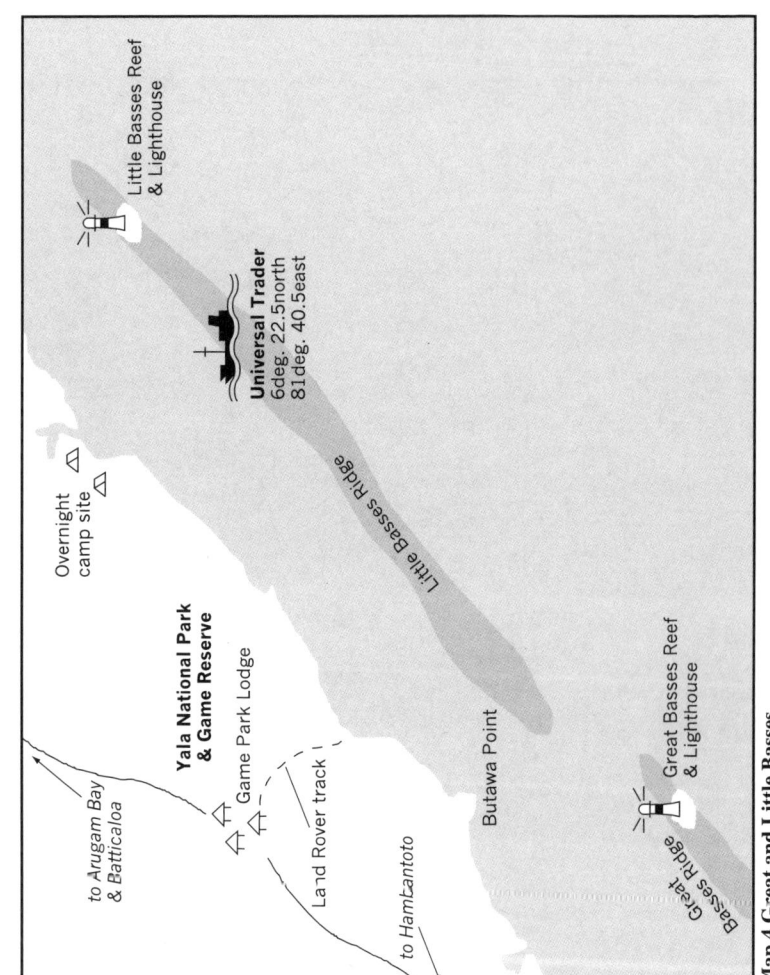

Map 4 Great and Little Basses

Little Basses Reef & Lighthouse

Universal Trader
6deg. 22.5north
81deg. 40.5east

Overnight camp site

Little Basses Ridge

Yala National Park & Game Reserve

Game Park Lodge

to Arugam Bay & Batticaloa

Land Rover track

to Hambantoto

Butawa Point

Great Basses Reef & Lighthouse

Great Basses Ridge

66

Five rounds rapid over the funnel, did the trick – the looters just vanished. We were left in total possession. Boarding was no problem as our departed friends had left a pilot ladder nicely rigged over the side.

Universal Trader had stranded on the Little Basses ridge in geographical position latitude 6 degrees 22.5 minutes north, longtitude 81 degrees 40.5 minutes east. She was in a very bad state. The hull had broken in two in line with the bridge front. The aft section, consisting of the engine-room, and the Nos. 4 and 5 holds, was sitting upright on the ridge more or less on the loaded draught.

The forward section, consisting of the Nos. 1, 2, and 3 holds, had partly slid off the ridge and swung round to port, with only the mast houses, derricks and vent cowls above water. The midship accommodation house was completely gutted by fire; nothing much for the looters there. The grain cargo in the Nos. 4 and 5 holds was still burning; the looters would not be able to salvage anything there either. The engine-room/No. 3 hold bulkhead, which was in way of the hull break, had collapsed. Standing on the engine-room upper gratings I could watch the seas rolling into the engine-room below me to smash against the aft bulkhead and foam around the top of the boilers and main-engine. Up in the burned-out bridge the engine-room telegraph stood at 'Finished With Engines'.

Every now and then the aft section of the hull would shudder as a larger-than-usual swell rolled into the engine-room. With regard to the extent of damage and prospects of salvage the situation was quite clear. *Universal Trader* was a 'Total Loss' in every sense: salvage was not an option.

While I was inspecting the casualty, my Lighthouse Service friends had been taking soundings to try and establish how far the stranded ship was from deep water. Considering her condition it was a rather academic exercise, but someone was bound to ask. The soundings showed that the casualty was sitting on the inshore edge of the ridge, which explained how the fore end had slipped partly off into deeper water.

There was no point lingering, as I had seen all I needed to see. The Little Basses Reef Lighthouse was 3.5 miles north-east of the casualty location so my friend, the lieutenant, took me over in the boat, the object being to see if the lighthouse-keeper could tell me anything about the stranding; hopefully, he might have seen something of interest. In view of the swell it was not considered safe to go alongside the little landing-stage at the foot of the lighthouse and the lieutenant and I transferred from the boat by means of a 'breeches buoy' type of system. The boat crew obviously used this method frequently and deployed the system with practised ease. In no time at all I was whisked over waves and rocks into the lighthouse.

This was the first occasion that I had been in a lighthouse. From the entry port, which was some 25 feet or so above the base, we climbed up a spiral staircase, rather like what one might find in a medieval castle, and arrived in the circular mess room where the keeper treated us to a pot of the finest Ceylon tea. After the lieutenant had explained the purpose of my visit the keeper told us what he knew about *Universal Trader*. The vessel had apparently run onto Little Basses during the night of 8-9 March. The first the lighthouse staff knew about the stranding was when they picked up a distress call from the ship. Shortly afterwards, they saw her silhouetted in the glow from the burning midship house. The crew, who had abandoned the burning vessel in the ship's lifeboats, were picked up at first light by a passing vessel, *Ocean Enterprise*, who had answered the distress call. This vessel was on a loaded voyage from Moji in Japan to Kuwait and, after passing details of the casualty to Colombo (see Lloyd's agent's message of 11 March), eventually landed the crew at Kuwait on 19 March. I never had any dealings with them, but the underwriters would have arranged for them to be interviewed at Kuwait by a surveyor, or solicitor.

What the lighthouse-keeper did tell me was that he had heard reports on his radio of an unidentified vessel being sighted close inshore at Batticaloa on 6 March and at Arugam Bay on 7 March. These locations were some 120 and 40 miles respectively to the north of Little Basses. This certainly sounded rather suspicious. Was this ship *Universal Trader* and, if so, what was she doing sailing southwards down the coast when she was bound to the north for Chittagong? Were they looking for a suitable location to arrange a phoney 'Total Loss' grounding? My investigative instincts were aroused and I decided that I had better get up to Arugam Bay and Batticaloa and talk to anybody who had seen the mystery unidentified vessel.

We had lingered too long at the lighthouse and the boat crew advised that there was not enough daylight left for us to reach the end of the track where the Land Rover had arranged to pick me up. What they proposed was to land on the beach immediately opposite the lighthouse and camp there. They had sufficient food and water and, in fact, often camped at this spot when they ran out of daylight during their runs to and from the lighthouse. It was a balmy tropical night. The seamen caught and cooked some crabs, which I think were the best I have ever tasted. The Ceylon crab, now of course known as the Sri Lankan crab, is a delicacy in all the best sea-food restaurants in the region – and these could not have been fresher.

The Lighthouse Service boat was equipped with a radio and we were therefore able to advise Willie, who was back at the game reserve chalet, that I was delayed but would be at the end of the track first thing in the morning.

Sitting on the beach that night I wondered why the ship had caught fire after she stranded. At first it seemed a bit suspicious, almost as if they were making sure the ship was a 'Total Loss'. Looking again at the message from *Ocean Enterprise* dated 11 March, I noted that they had reported wrinkles and cracks developing on the port and starboard shell plating, followed by leakage of fuel into the engine-room from the settling-tanks. Thinking about it, these wrinkles and cracks they mentioned were almost certainly in way of the engine-room forward bulkhead where the hull later broke in two. The forward ends of both settling-tanks are in this area and they also probably started to fracture at the same time as the hull and the engine-room bulkhead. A fracture on the inboard bulkhead of a nearly full settling-tank could have caused fuel not just to leak, but to spray into the engine-room, probably into contact with a hot section of the adjacent boiler. The boilers themselves may have been affected by stresses to the ship's structure and some of the hot furnace brickwork may well have been exposed. My conclusion was that there was nothing sinister about the engine-room fire; it was in all probability, a consequence of the stranding and subsequent break up of the hull. But why the ship had been so close inshore in a dangerous area, clearly marked by two lighthouses, was a mystery. With regard to Little Basses ridge the Admiralty sailing directions state:-

It is not advisable for any vessel, except those of light draught, to attempt to cross any part of the ridge.

Universal Trader in her fully loaded condition would probably have had a deep draught of more than 28 feet making it essential to keep well clear of the Little Basses ridge.

By 10.00a.m. next day we were on our way to Arugam Bay to see what we could find out about the mysterious ship seen just off shore on 7 March. Before setting off I sent

70

a message to the underwriters, describing the condition of the casualty. This was published in part in Lloyd's List of 22 March:-

Colombo 21 March

Steamer Universal Trader *broken in two at bridge front and fore end submerged in 50–60 feet of water. Aft section firmly aground in approximately 20 feet water with midship house completely gutted. Cargo in No. 4 and 5 holds burning fiercely and deck and shell plating in way badly buckled. Engine and boiler room open to sea and flooded above level of engines and boilers. From soundings taken vessel is on the inshore edge of the reef.*

Salvage Association Special Surveyor

I had also advised London of the nature of the information obtained from the lighthouse keeper, and that I was proceeding north to investigate. As this was confidential information for the underwriters alone, it does not appear with the rest of my message.

We arrived at Arugam Bay early in the afternoon and spent the rest of the day tracking down the people who had reported seeing the ship. On questioning them it soon became clear that we were not going to get a positive identification, as the evidence we received was rather confused. Some said they had seen a large ship, others described what appeared to be a small coastal vessel. I suppose we spoke to about ten people who claimed to have seen a vessel close inshore, and received 10 different descriptions. At the end of the day we had no evidence of *Universal Trader* having been close inshore at Arugam Bay on 7 March.

That night we spent in a small local guest-house. All that

I can remember of Arugam Bay is miles of golden sandy beach, a most uncomfortable bed in the guest-house and a fan that operated only at full speed. You either lay and sweated with the fan off, or froze in a howling gale with the fan on.

Next day we drove up the coast to Batticaloa. My colleague from Aitken Spence & Co. Colombo, who had proved invaluable in tracking down the various witnesses at Arugam Bay, knew the harbour master at Batticaloa and we were therefore fairly confident that we would now get hold of some positive information.

We certainly did get very positive information. The harbour master had seen the ship himself, quite close to the port on 6 March. After about an hour or two it had sailed off southwards. His description was quite clear: it had been a coastal vessel of about 1,000 gross tons, having two hatches and the bridge and accommodation aft. Whoever she was, she was not our *Universal Trader*.

Well, that was that. My hopes of cracking a case of maritime fraud, with visions of being congratulated by the chairman of Lloyd's, evaporated into the thin, tropical air. There was nothing else for it but to drive back to Colombo and advise the underwriters of the bad news.

Our return journey was through some of the most beautiful scenery in the island. We left Batticaloa and proceeded south-west along route A5. Slowly the central mountains rose higher and higher in front of us and, gradually, the hot humid air of the plains along the eastern seaboard gave way to a cooler, crisper atmosphere. My colleague, Willie Vandenberg, told me that this route to Colombo, which would take us through the hill resort of Nurelia (now Nuwara Eliya) at over 6,000 feet, was one of the most beautiful road journeys in Ceylon. It certainly lived up to his description. We had planned to spend the night at Nurelia, but we ran out of daylight and had to stay at Bandarawela instead. But it is in the high country at 4,500 feet, in the middle of the tea plantations, and so was pleasantly cool.

Next day we travelled on to Nurelia which, for some reason or other, reminded me of Pitlochry in Perthshire. No doubt it was the cool air – almost cold in comparison to the coastal region – the hills, the pine trees and the flowers. We stopped at a market, where Willie bought fresh fruit and vegetables for his family in Colombo, then journeyed on, slowly losing height as we descended through the pleasant, tea country towards the steamy heat of Colombo, where we arrived about 5.00p.m.

It had been a most interesting trip around the southern half of Ceylon, the sort of safari-like expedition that would cost today's tourists thousands of pounds – and I got it all absolutely free. Furthermore, I was paid to do it! Now what more could you ask for?

4

Rangoon and Akyab

Akyab is the main port and administrative centre of the Arakan region of Burma. It lies on the eastern side of Akyab Island at latitude 20 degrees 08 minutes north, longitude 92 degrees 54 minutes east, some 350 miles north-west of Rangoon. The island lies in the delta area formed by the Mayu and Kaladan rivers. The wet, forested hills of the Arakan interior are sparsely populated and most of the population is concentrated in the delta area and on Akyab Island. As one of the main ports of Burma, Akyab handles both local and international trade. The main export cargo is rice and this is cultivated both on the island, where there are quite a few rice mills, and in the adjacent delta area. Ocean-going vessels normally load rice from barges in the anchorage at the mouth of the Kaladan River. As it passes Akyab, the river is divided into two branches by the islands of Ngapi Kyun and Paw Kyun. The western branch of the Kaladan flows to the west of these islands, which form the left bank of the river. The coast of Akyab Island and the town itself form the right-hand bank of the river.

To facilitate cargo work, ocean-going ships anchor in the western branch of the river close inshore between the main wharf and the mouth of the Satyogya *chaung* (Burmese for stream), about two miles upstream. This anchorage is reasonably sheltered from the effects of the south-west monsoon, but offers little protection from the severe tropical cyclones

74

which, every three or four years, devastate the Arakan and Bangladesh coast during the months of April and May. The year 1968 was to be one of those years.

The Greek flag steamer *Gero Michalos* arrived at the Akyab anchorage in ballast from Trincomalee on 27 April 1968 and, next day, commenced loading bagged rice for Colombo. *Gero Michalos* was owned by N. Michalos & Sons Maritime Co. Ltd., of Chios, with Victoria Steamship Co., London, as managers. The vessel was built in 1946 at Ailsa Ship Building Co. Ltd., Troon, Scotland, for the Ministry of War Transport, and launched as *Empire Warner*. Of just under 3,000 gross tons she was propelled by a 1,100 HP triple-expansion reciprocating steam engine, giving a service speed of 9.5 knots. N.Michalos & Sons had acquired the vessel in 1965.

On 7–8 May 1968 a severe tropical depression in the Bay of Bengal deepened and moved north towards the Arakan coast. By 9 May it had been classified as a cyclone. At this stage *Gero Michalos* had almost completed loading cargo, but the approach of the cyclone caused all cargo work to be suspended. The vessel was closed up and engines put on 'stand-by' as the approach of the cyclone was awaited. In the early hours of 10 May the cyclone struck Akyab, causing extensive damage to both the town and the port. Despite the precautions taken on the Greek steamer, the force of the cyclone initially caused the vessel to drag anchor and ground on top of an old wreck close to the Akyab water-front. This resulted in damage to the underwater hull, causing serious ingress of water in way of the engine-room and one of the cargo holds. As the cyclone continued, *Gero Michalos* was driven clear of the sunken wreck and across the anchorage to eastern side of the fairway. The pumps were unable to cope with the flooding situation in the engine-room and, by noon on 10 May, power was lost as the rising water-level reached the boiler furnaces and the fires had to

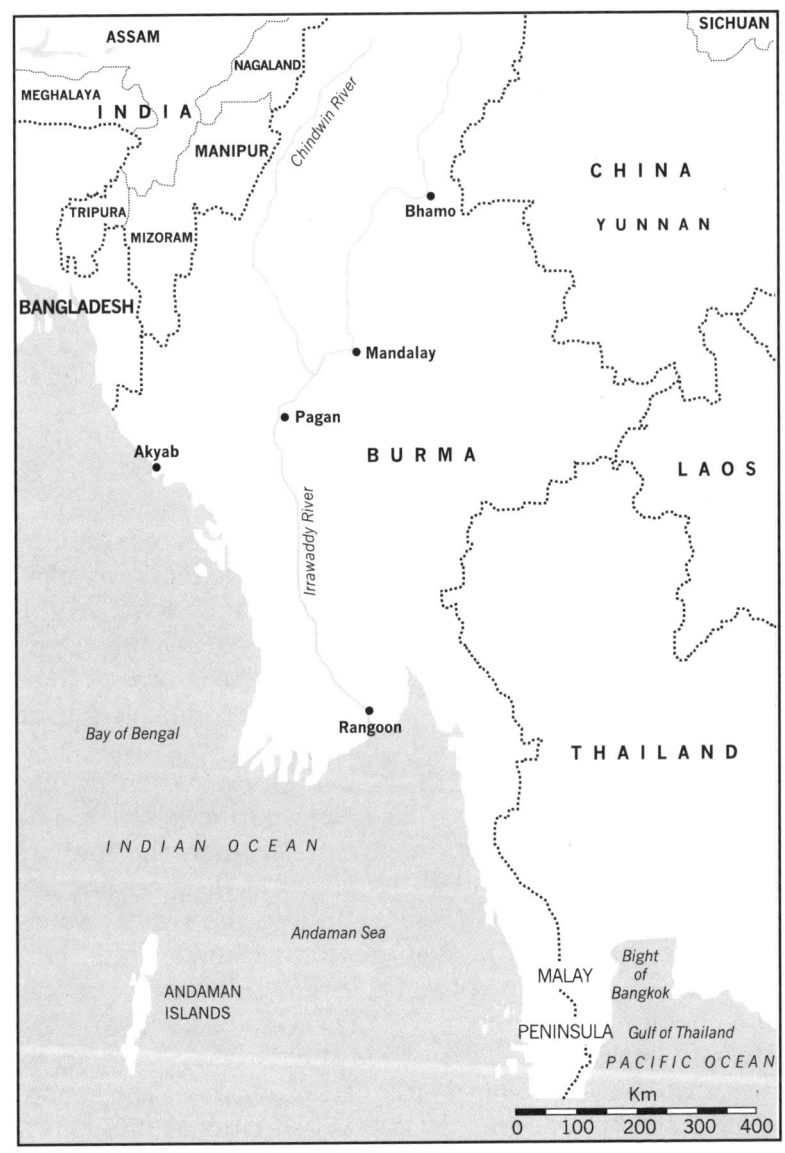

Map 5 Burma (Myanmar) and neighbouring countries

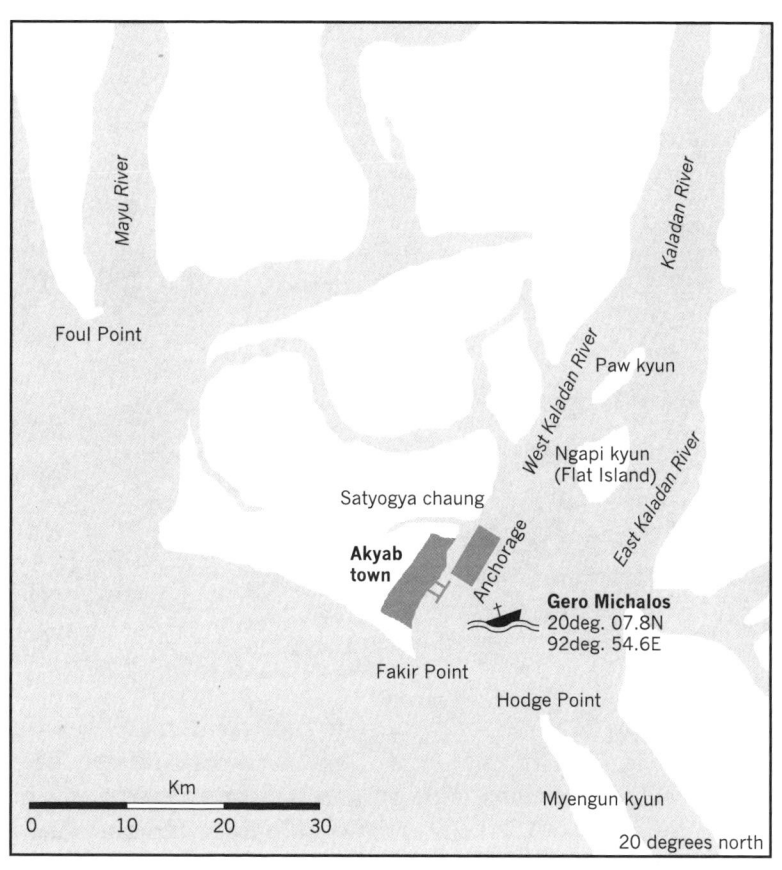

Map 6 Akyab Harbour and Kaladan River

be extinguished. The vessel then progressively flooded throughout and settled on to the bottom with only the forecastle, bridge deck and poop remaining above water.

On being advised of the situation the ship's managers arranged for the crew to be repatriated to Greece and, at the same time, informed their insurance brokers of a potential 'Total Loss' claim on the Hull & Machinery Underwriters. The outcome of all this was that Salvage Association, acting on behalf of the underwriters, requested Ritchie & Bisset to send a surveyor to Akyab to ascertain the prospects of refloating and repairing the casualty within the insured value.

As Ritchie & Bisset's Burma 'expert,' I presented myself at the Burmese Embassy in Singapore on 15 May and completed, in quadruplicate, the application for a visa to visit Burma. The reason for the visit was clearly stated as: 'To inspect the cyclone-damaged Greek steamer *Gero Michalos* at Akyab on behalf of London underwriters.'

My previous visa application in 1967 had taken one week so, having completed the formalities at the Embassy, I returned to the office and continued with my routine Singapore survey work. After a week had passed I spoke to the Embassy about the status of my visa only to be told that they had not yet received any word from Rangoon. This did not worry me particularly as I had plenty of work in Singapore to be going on with. The underwriters in London, however, were becoming anxious for news and we kept getting telex messages enquiring as to when I was going to arrive at Akyab. After another week had come and gone I called at the Embassy to try and find out what was going on. The answer was not long in coming. 'Your application for a Burmese visa has not been approved by Rangoon,' I was told. No reason was given.

The underwriters, on being advised of this unusual turn of events, told us to re-apply while, at the same time, they would approach the Burmese Embassy in London and explain the need for me to get to Akyab as soon as possible. Back I went to the Embassy, which was then located in a lovely old Singapore 'Black and White' house, just off Tanglin Road. Another visa application form was completed in quadruplicate, together with goodness knows how many photographs. Another week or so passed by with no news from the Embassy. Then suddenly, on 18 June, just when I had began to think that the whole thing was a non-starter, the Embassy phoned the office and told me that my application had been approved.

Friday 21 June saw me, complete with visa in passport, out at Paya Lebar airport to board a flight to Bangkok. Once there, I caught a Union of Burma Airline flight to Rangoon. On arrival I was met by my friend, U Shwe Yee, who was the local Lloyd's agent and had been a great help during my previous visit to Burma in November 1967. He was going to have his work cut out this time. My intentions were to arrange the trip to Akyab on the Saturday and then fly up on either Sunday or Monday, depending on the availability of flights. I explained this to U Shwe Yee, only to be told that it was not quite so simple.

'I have to get you a permit from the Ministry of Home Affairs before you can go to Akyab,' he explained.

'What are you talking about? It took them nearly five weeks to give me a visa and it specifically stated on the application that the reason for the visa was to inspect the cyclone-damaged Greek ship at Akyab. So what possible problem can there be now?'

'I am afraid the law is that, despite having a visa, a foreigner cannot leave the city limits of Rangoon without permission from the Ministry of Home Affairs. We must, therefore, get a permit for you before you can fly up to Akyab,' my Burmese friend replied. 'However, it should not be a great problem. I will call at the Ministry tomorrow morning and – you never know – maybe you will be able to fly up on Sunday or Monday'.

Next morning, after glancing through the local Rangoon paper, *The Nation*, I set off for the Lloyd's agent's office in Merchant Street, a ten-minute walk from the Strand Hotel. The news was not good. U Shwe Yee had been to see the powers that be at the Ministry and had simply been told to come back on Tuesday. So that was that for the next four days. The earliest I was going to get to Akyab would be Wednesday, and that was by no means certain. I had brought some of my outstanding paperwork with me from Singapore but that only kept me occupied until midday on the Sunday. From then until Tuesday morning I just had to twiddle my thumbs. There was very little to do in Rangoon in those days.

Ten years previously I had been 2nd engineer on the Henderson Line's *Bhamo*. In those days I had spent three or four weeks at a time in Rangoon while the ship discharged cargo at the Sule Pagoda wharf and then, out at a double-mooring berth in the river, loaded a full homeward cargo from barges. I have described elsewhere how, in our spare time, we played football against other ships and local teams, visited the famous Shwedagon Pagoda and went sailing on the Inya Lake. I had enjoyed Rangoon, for the ordinary Burmese, despite the apparent idiosyncrasies of the various departments of General Ne Win's Government, are a friendly, intelligent and charming people; in fact, our Burmese crews on Paddy Henderson's ships were by far the best I ever sailed with. But now I was stuck in an almost empty Strand Hotel with nothing much to do and no old shipmates to yarn to: no Colin Kerr to entertain me with amusing tales of his long years of service in some very hot – and some not so hot – stations.

Tuesday finally arrived and, after giving U Shwe Yee time to see the officials at the Home Affairs Ministry, I hotfooted it round to his office in Merchant Street to find out my fate.

'Well, what did they say today?' I asked.

'They told me that they would let me know, so we will just have to wait and see what they decide,' was his disappointing news. It was a case of: 'Don't call us we'll call you.'

I could not believe it. Having, after much trouble, obtained a visa to visit Burma in order inspect *Gero Michalos* at Akyab, I was now being prevented from travelling from Rangoon to Akyab for some reason known only to 'the men from the ministry'. I asked U Shwe Yee if he had any idea why the authorities were being so difficult. On my previous visit in November there had been no problem arranging a flight down to Great Coco Island, and *that* was a restricted military area. The answer appeared to be that on that occasion the Burmese Government had had a direct financial involvement in the rice cargo in the ship aground on the island and were anxious to find out if any of it could be saved. With regard to the ship at Akyab, we could only assume that they had already been paid for the cargo in *Gero Michalos* and were, therefore, no longer interested in the matter.

Of course I should not really have been surprised by the problems with the authorities, for this was the time when General Ne Win and his Revolutionary Council were in government, their official policy having the grand title of *The Burmese Way To Socialism*. In addition to being a ruthless dictator Ne Win was extremely superstitious and was fascinated with the 'science' of numerology – his lucky number was nine. No important decisions were made that were not somehow linked with this number. Foreign businesses, including banks, were nationalised and contact with foreigners severely restricted. Under his despotic and chaotic rule one of South East Asia's potentially richest countries had become one of the poorest. With that sort of person in charge of the country it was small wonder we were not getting much joy from the Ministry of Home Affairs as far as travelling to Akyab was concerned.

The only thing to do was to inform London of the situation and let them decide whether I should stay any longer or return to Singapore. I was inclined to favour the latter choice. However, London's reply was that I should remain in Rangoon and continue to use my best endeavours to get the necessary clearance to enable me to travel to Akyab. In the meantime, they gave me another job to keep me gainfully occupied.

The Inland Water Transport Board (successor to the old Irrawaddy Flotilla Company) had recently taken delivery of a new tug built by a Dutch shipyard. The IWTB were alleging that the tug's dead-weight capacity was less than that stated in the builders' specification. The contract between the builders and the owner apparently contained a penalty clause concerning any such discrepancy. What the importance of dead-weight capacity was in the case of a river tug I never did find out. Perhaps the tug had some cargo-carrying capability. Anyway, the problem for me was that the owners, IWTB, were claiming compensation from the builders in accordance with the penalty clause. The builders had therefore notified their Builders' Liability underwriters of a possible claim. My new assignment was to investigate the situation and ascertain whether there was in fact a discrepancy in dead-weight that would be covered by the penalty clause in the builders' contract and, if so, what would be the fair and reasonable quantum of the claim. My contact at IWTB was given as the Comptroller of Stores.

U Shwe Yee contacted IWTB on my behalf and arranged a meeting with the comptroller of stores for Friday morning. Full of enthusiasm, after my week of inactivity, I was round at their office at 50 Phayre Street promptly at 9.00a.m. and was shown into the comptroller's office on the ground floor.

He was a wee gentleman wearing the traditional Burmese longyi. The floor of his office was covered with files, piled up to a height of three or four feet, and I had difficulty picking my way through them to reach a chair. He gazed at me rather suspiciously over the top of his half-moon glasses while I explained why I had come to see him. It soon became apparent that he had no idea what I was talking about and, as soon as politeness allowed, he ushered me upstairs to the third floor, to an office of someone in an even more senior position. The title on the door read: 'Officer on Special Duty'. I was impressed. At this rate, I thought to myself, I should be meeting General Ne Win himself by the middle of the afternoon.

The 'Officer on Special Duty' turned out to be a lieutenant commander in the Burmese Navy who occupied a very large office indeed – a far cry from the pokey, file-filled den of the comptroller of stores, three floors below. At this time in Burma all civilian government departments included an 'Officer on Special Duty' who was a military officer and, as such, a representative of General Ne Win's Government. His role was akin to that of a political commissar.

The lieutenant commander knew all about the tug and he explained to me that they had taken very precise measurements and their subsequent calculations showed that the tug's dead-weight capacity was x tons short of the figure in the builders' specification. They would, therefore, apply the penalty clause and claim the appropriate costs from the builders.

After describing the scope of my involvement I suggested that they allow me to visit the tug and check their measurements and calculations. If I found that the tug's dead-weight was, in fact, deficient by the amount they were alleging, then I would advise the underwriters concerned accordingly.

'I am afraid that you cannot visit the tug,' I was told. 'It is busy working on the Irrawaddy River beyond Mandalay and it would not be convenient for you to go aboard.' That was disappointing news. For a moment or two I had thought that, at last, my long-held ambition to visit Mandalay was about to be realised.

'Well, if I can't visit the tug,' I replied, 'maybe you could show me your calculations. You never know – perhaps, if I report that your calculations appear to be accurate the underwriters might accept the claim, or at least reach a settlement with the builders. But I will really need to provide some sort of factual evidence for them to be able to reach a settlement acceptable to all parties.'

'Oh, I am afraid we could not let you see our calculations,' I was told.

In spite of explaining as best I could that neither the builders nor the underwriters were likely to accept their loss of dead-weight claim without some sort of supporting evidence from an independent party – in other words me – I was unable to make any progress. The lieutenant commander, however, did promise to let me know if there was any change to their position; with that I was shown to the door.

Back I went to U Shwe Yee's office where I sent off a telegram to London explaining the result of my visit to the Inland Water Transport Board. I had now been in Rangoon for one week and achieved absolutely nothing. A telephone call from U Shwe Yee to the Home Affairs Ministry indicated that nothing had changed concerning Akyab, so it was back to the Strand Hotel for me, to read my copy of *The Nation* and await developments.

The Strand Hotel, in those days, was a fairly quiet sort of place. There were very few guests and for most of the time I was the only European. In fact, the only non-Burmese in Rangoon appeared to be the Embassy staffs. The only beer was the local Mandalay Beer. This is not a bad brew but it tended to go a bit flat and tasteless if it was too cold. To combat this tendency I got the barman to take a bottle of

Mandalay beer out of his fridge about 6.00p.m. so that it would be reasonably lively when poured out for me at 7.00p.m. At about that time the hotel orchestra, a rather elderly lady pianist and a young man who played the violin, would commence their evening's entertainment. The Palm Court Orchestra (Sunday evenings on BBC radio) they were not. I never really recognised anything they played. In fact, much of the time I suspected they each played different tunes. One evening after dinner I spoke to them and asked if they knew any Scottish tunes.

'Oh yes – we will play one for you tomorrow night,' they assured me.

Next evening I entered the lounge and, with a 'One, two, three . . .' they burst into 'When Irish Eyes Are Smiling'. Well, never mind – at least they were both playing the same tune and I recognised it right away.

As mentioned before, I was for most of my stay the only European in the hotel. Then, one morning at breakfast, towards the end of my second week in the hotel, I heard the sounds of hearty laughter in the dining room. I looked over to see the source of such unusual merriment and saw a large, slightly rotund individual joking with one of the waiters. He looked decidedly English and this, indeed, proved to be the case. We got chatting after breakfast and I discovered he was Tommy White, a service engineer from Perkins of Peterborough. The Burmese Military had numerous Perkins diesel engines in their heavy vehicles, generating units and floating craft and, once a year, they discussed their spare-part requirements with one of the Perkins engineers. In view of the fact that he was in Burma at the specific invitation of the Military Authorities, he faced none of the problems that I was experiencing. He could not understand why I should be having any difficulties at all and was full of advice as to how I should deal with Burmese officialdom.

One afternoon about two days later, I bumped into Tommy in the hotel lobby. He told me he was going to visit the Anglican Church to try and obtain an old 1938 or '39 baptismal record for a friend of his back in Peterborough. Did I want to come for the walk? I certainly did; so off we went to St Paul's Anglican Church, only some 20 minutes from the hotel. Unfortunately, our visit was in vain. The padre told us that all the pre-1942 records had been removed prior to the arrival of the Japanese Army, but had subsequently been abandoned and lost somewhere in Upper Burma.

Next afternoon, the two of us went for another walk round the town and paid a visit to the Catholic Cathedral. It was the first time I had been inside a Catholic church and I remember being impressed by the carvings round the walls and the numerous Burmese men and women sitting here and there, praying silently or just quietly meditating.

On the following day I studied a large map of Rangoon hanging on a wall beside the reception desk and spotted a church marked 'Church of Scotland'. So, after Tommy returned from his day's work sorting out the engine spares, we went off on another 'church safari.' However, on reaching our destination that evening we found the doors closed and locked. I knocked on the door and it was eventually opened by an Indian gentleman who turned out to be the caretaker. The Church of Scotland, he informed us, had been gone from these parts for several years. He invited us in to look round the church, which was now in the hands of the Burmese Christian Association and the Baptists. Inside, the two religious groups were kept apart by a pierced steel plank (PSP) barricade, which neatly divided the church in two. During the war in Burma, PSP was used by General Slim's 14th Army to build airstrips, roads, and a thousand other things. After the war the Burmese used it for nearly everything you could think of. There was nowhere you could go in Burma without seeing miles and miles of PSP.

Next day was Tommy's last day in Rangoon and, after breakfast, he set off cheerfully for an appointment with the Burmese Army at one of their vehicle maintenance depots – the equivalent of a British REME workshop. I wandered round to Merchant Street to have my morning coffee with U Shwe Yee, check on the latest situation regarding Akyab and see if there was any news from London. There was no movement in the Akyab situation. Regarding the problem of the tug, London asked that I keep in touch with Inland Water Transport in case they changed their mind about providing access to it. Back I went to the hotel to find, much to my surprise, that Tommy had arrived there before me. He was jumping up and down and spluttering with rage: most certainly not his usual cheerful self.

What had happened was, that after he had finished working on their spare-part requirements, he had suggested to the officer in charge that he often found it useful to collect a dozen or so of the mechanics and foremen round one of the partly dismantled Perkins engines. He would then give a short talk about some of the common problems encountered in service and answer any questions that they might have. Quite a good idea agreed the Burmese officer but, unfortunately, all the engines are in the workshop area and foreigners are not permitted to enter.

'Well, that's that then,' said my friend and prepared to take his leave.

'Now, before you go,' said the Burman. 'perhaps you could help us with a problem we have with one of your engines. It has suffered complete seizure on two occasions and we cannot find the cause.'

'OK, that's no problem,' said my friend. 'Let's have a good look at it.'

'Oh, you won't be able to see it – it's in the workshop,' explained the officer.

I am afraid I could not help laughing, but it was a relief to realise that I was not the only one who had problems with officialdom. Anyway, my friend from Peterborough now had a better understanding of my difficulties in organising the trip to Akyab and in dealing with the Inland Water Transport Board.

Some time during my third week in Rangoon, maybe about 8 July, the agents for the vessel up at Akyab received a cable from the ship managers in London, who were obviously getting a bit fed up at the lack of progress with their potential 'Total Loss' claim. It was a classic Greek owner's message and I quote it below:-

Gero Michalos fail to understand why Walker being prevented from proceeding Akyab as he is on legitimate business stop Consider attitude of Burmese Authorities to be unwarranted obstruction stop Kindly revert soonest with latest status stop Victoria Steamship London

U Shwe Yee examined the message closely, then told me that he would take it round to the Home Affairs Ministry and show it to them. Well I thought, they will either send me up to Akyab, throw me in jail, or kick me out of the country. In fact, nothing at all happened for a day or two; then, on or about, 11 July U Shwe Yee, on making his umpteenth visit to the Ministry, was told, 'The foreigner can now go to Akyab'.

He phoned me at the Strand with the news and I dashed round to see him.

'Where's my permit?' I asked.

'They said you don't need a written permit; apparently it's now in order for you to just fly up to Akyab,' he assured me.

So, after three weeks of sitting about doing nothing, I flew up to Akyab on Friday 12 July on a Union of Burma Airline Fokker Friendship, arriving around the middle of the day. U Shwe Yee had a sub-agent in Akyab by the name of Yusoff. He was a Muslim whose family had come originally from East Pakistan or Bengal. It had been arranged that he would meet me and fix up some place for me to stay. Mind you, there was not much choice: all that was available was a travellers' guest-house. In colonial days it had been the harbour master's bungalow. Now it was in a state of considerable disrepair. There was no running water; only a well at the bottom of the garden. The garden itself was badly overgrown and to reach the well you had to make your way through a veritable jungle of head-high elephant grass, vines and creepers.

In the afternoon Yusoff picked me up in his jeep, another bit of former 14th Army equipment, and took me down to the wharf where he had organised a boat to take us out to *Gero Michalos.*

The casualty lay upright, about one mile offshore, at the eastern side of the west river fairway, in about 35 feet of water. Only the accommodation house, forecastle and poop were above water. At low tide the hatch-coamings just broke the surface. It was low water when I reached the sunken vessel and there must have been about twenty or so sampans clustered round the hatch-coamings, with their occupants busy diving in and out of the holds.

In answer to my enquiry as to what they were doing, Yusoff explained that they were removing the bags of rice which they would then take ashore, wash in fresh water, and then spread the rice out to dry in the sun. Following this treatment the rice could be used in the normal way. Rice is the staple diet in the Arakan, as it is in most of the region, but the level of poverty in this part of Burma meant that it was like gold to them. In any event, the water at the casualty location was brackish, owing to the Kaladan River flowing into the anchorage, so restoring the rice to an edible condition was not really a problem. However, the actual removal of the rice bags from the hold was anything but easy. Yusoff told me that at least three of the locals had lost their lives as a result of being trapped under water in the hold wings.

An inspection of the midship house showed that the accommodation spaces and bridge had been completely gutted, presumably by the same gentry who were now busy trying to remove the rice cargo. Only the bare steel shell remained: every single item of navigational and radio equipment had been removed together with all the accommodation fittings, right down to the last nut and bolt. Up forward on the forecastle the windlass had been dismantled and most of it removed. At the time of my inspection the main drum/gypsy shaft was sitting next to the ship's side, presumably waiting for a sampan to come along and pick it up on the next tide.

Refloating would not present a major problem to a competent salvor but the cost of the operation would be substantial as all the salvor's equipment would have to be mobilised from Singapore. However, repairs to the flooding damage, the replacement of the entire accommodation space fittings and furnishings, plus the renewal of all the navigational and radio equipment, would exceed the insured value, without even considering the salvage costs. All in all, salvage and repair of this old vessel would not be economically viable.

There was nothing else to inspect so it was back ashore with Yusoff, who then gave me a lift to the guest-house. The house-boy who was supposed to look after the guests had rather odd-looking hands. After a while it dawned on me that each hand had five fingers and a thumb! Too late I realised that there was no means of cooking anything and I had to make do that night with a pot of tea and a couple of biscuits that I had brought with me. Boiling the kettle for the tea was a major production. There was no stove in the bungalow so the 'Boy' and I had to light a fire outside. It had been raining that day and the wood was wet and reluctant to burn. Eventually, however, the kettle reached the boil and, at long last, tea was served.

Next morning Yusoff took me round to the airline office to see about a flight back to Rangoon. To my utter horror I was told that the next flight would be on Monday afternoon. I was going to have to spend all of Saturday and Sunday, not to mention most of Monday, in the travellers' guest-house. It was quite a thought. I explained the cooking problems to Yusoff and he promptly shot off back to his office and returned with a paraffin camping stove and a container of paraffin. It was just what was needed. I had been warned that conditions at Akyab might be a little bit primitive and had brought with me some packets of army boil-in-the-bag MREs (meals ready to eat); these were easily heated up on Yusoff's little stove.

The three days I spent at Akyab turned out to be not so bad really, thanks mainly to friend Yusoff's stove and a couple of books he lent me to help pass the time. On the Saturday afternoon I was joined at the guest-house by a young Burmese doctor and his father; they had been visiting friends near Akyab. It was shortly after Professor Christian Barnard performed the first ever heart transplant operation at Cape Town. I can remember the doctor telling me again and again how this was the most amazing medical feat that had ever been carried out. On the Sunday morning Yusoff turned up with his jeep and took me on a grand tour of Akyab Island and I saw some of the damage the cyclone had caused. One of the rice mills had taken a direct hit and looked to be a complete wreck.

Yusoff called again on the Monday morning and, after I

had handed over his stove and the books he had kindly lent me, I asked him what time we should go out to the airfield.

'Well,' he said, 'we usually wait until the aircraft flies over. That gives us plenty time to get out to the airfield.'

'What time is it scheduled to depart?' I asked.

The answer to that was 'about 2.00p.m.', so I suggested that we get to the airfield at least one hour before. No way was I going to miss that flight; goodness knows when the next one would be. Yusoff thought that 1.00p.m. was a bit early. But if that's what I wanted to do he would pick me up at about 12.45p.m.

We arrived at the airfield promptly at 1.00p.m. and for the next hour and a half absolutely nothing happened in the way of aircraft arrivals or departures. I can remember seeing Yusoff in conversation with the airfield manager at about 2.30p.m.

'What did he have to say?' I asked, after they had finished talking.

'He was just saying that the aircraft has not yet left Rangoon.'

Obviously the local practice of waiting until the aircraft flew over Akyab before setting out for the airfield was based on years of sound, practical experience.

The aircraft eventually arrived at around 4.00p.m. After saying 'good-bye' to Yusoff, without whose assistance my stay would have been very uncomfortable indeed, it was back to Rangoon where good old U Shwe Yee was waiting to take me back to the Strand, where I had my first real wash in four days. After the guest-house at Akyab, the Strand seemed to epitomise the last word in luxury.

First thing next morning I sent the following telegram to London regarding the situation at Akyab:-

Rangoon 16 July 1968

Steamer Gero Michalos *lies upright in some 35 feet of water in Akyab fairway at geographical position 20 degrees 07.8 north 92 degrees 54.6 east stop At low water main deck submerged but forecastle bridge deck and poop above water stop Accommodation spaces together with radio room and wheelhouse completely gutted by looters who now busy removing cargo stop For Under-writers' guidance cost of salvage and repair will exceed insured value stop*

Walker/Lloyd's Agents

The next task was to see if any progress could be made with the Inland Water Transport Board regarding the dispute about the dead-weight of their new tug. The answer to that was 'no'. They did not want to discuss the matter with me any further. So, on 17 or 18 July, I sent a cable to London saying that I saw no prospects of obtaining any co-operation from the Inland Water Transport Board and asking if I could return to Singapore. London replied in the affirmative. They had clearly arrived at the fairly obvious conclusion that they were not going to get any more mileage out of me at Rangoon.

Friday 19 July was a public holiday but the British Overseas Airways office was open so I strolled round from the Strand to make sure I was booked on the Saturday flight to Singapore. Being a holiday the streets were very quiet, with very few people to be seen. On my way back from the airline office I passed a group of trishaw riders at the corner of Strand Road and Phayre Street, they always used to hang about there offering to take you to the Shwedagon Pagoda,

93

the Inya Lake, or wherever. On this occasion though, one of them came right across the pavement to get in front of me. I thought that a bit unusual but was not really paying much attention to him. I became aware, however, that he was pointing at something and that the pointing hand was badly disfigured. At that stage an alarm bell started to ring in my brain; at the same time I heard a metallic clink on the pavement near my feet. I looked down just in time to see the trishaw rider placing one big, bare foot on top of my nice, sterling-silver Parker pen. It had been in the breast pocket of my shirt, together with my airline ticket, and the villain had removed it, without me feeling a thing, while I was looking to see what he was pointing at. Unfortunately for him he dropped it and, by giving him a bit of a push, I was able to recover it. The trishaw rider, presumably cursing his bad luck at letting a small fortune literally slip through his fingers, scurried back to rejoin his colleagues, while I retired to the safety of the Strand Hotel.

On Saturday 20 July I said goodbye to my good friend, U Shwe Yee, at Mingaladon airport and left Rangoon to return to the commercial hurly-burly of Singapore. I had been in Burma for four weeks and had achieved what would normally have been accomplished elsewhere in about two days. In retrospect I do not really think that there was any deliberate policy to obstruct my journey to Akyab. It was probably just that nobody in the Ministry of Home Affairs was prepared to be the person to say, 'Yes, it's OK for the foreigner to travel to Akyab'. The whole situation was just a by-product of the political climate of those years. Nowadays, the present government in Burma gets a very bad press in the western media; but in my opinion they have, despite their faults, greatly improved the overall efficiency of the country. When I read some of these media reports I think to myself, 'It's all very well for you to pontificate about the situation in Burma, chum, but you should have been there with me in 1968. Now that would have given you something to whinge about.'

In Singapore I was, for many years, an enthusiastic member of the Hash House Harriers, a Monday evening cross-country running group, whose athletic efforts were invariably followed by the consumption of a fair amount of liquid refreshment. In 1968 I was one of the joint masters responsible for organising the runs, the weekly newsletter, etc. Before leaving Singapore I had told them I was off to Burma for about three or four days but would be back for the run on the following Monday. That was now more than three weeks ago. So, to let them know that I was still alive and kicking, one evening in the Strand Hotel I scribbled down a poem – a bit of a parody on Kipling's 'Road to Mandalay' – then posted it off to Singapore for inclusion in the newsletter.

I have finished this short account of my time at Rangoon and Akyab with the poem as it appeared in the newsletter of 22 July 1968. The 'young Malcolm' referred to in the penultimate line was the treasurer, who had the unenviable task of collecting the monthly subs.

IN MEMORIAM

For S. Walker, presently a 'prisoner' of the Revolutionary Government of the Union of Burma; one of our nobler casualties and a current Joint Master.

———————

By the Shwedagon Pagoda, gulping down a glass of
    tea,
There's an old Joint Master sitting, and he thinks of
    naught but thee;
For the Shandy's in the bucket, and he hears the
    Hounds all say:

95

'Come ye back, ye daft old Harrier; come ye back
    from Mandalay.'

Oh he's learning here in Burma what the oldest
    Hashman tells:
'Once you've heard the Hash a-calling you'll heed no
    other yells.'
For the wind is in the palm trees and the kampong
    folk all say:
'Come ye back, ye daft old Harrier; come ye back
    from Mandalay.'

On the road to darkest Akyab you can see the Master
    play;
'As he runs and checks and cries "On! On!" 'tween
    Rangoon and Mandalay.
For the Temple Bells are ringing, and he hears young
    Malcolm say:
'Come ye back, ye miserable b*st*rd – you've last
    month's sub to pay!'

5

Chief Lemon Tree

One day in July 1972 as I sat in the Salvage Association's Southampton office, toiling away at the usual surveyor's paperwork, my boss, Clive Key, strolled into my room and casually remarked that we had been given a new assignment. Apparently, the freezer trawler *Nagroor III* had run aground down in West Africa, at some place called Sherbro Island.

'Could you pop down and have a look at her, ascertain the extent of damage and prospects of salvage?' he asked. 'London don't seem to know much about the place, apart from the fact that it is somewhere in West Africa.'

In the depths of my memory something stirred. 'Sherbro Island . . . Sherbro Island,' I mused. Suddenly, it clicked. 'I know where that is – I've been there.'

'In that case, you're just the chap we need!' enthused old Clive.

My thoughts travelled back some 16 years to 1956 when I was serving as 2nd engineer on the Henderson Line cargo ship *Kaladan*. As I have described elsewhere, Paddy Henderson was originally engaged in the Burma trade; hence the Burmese names of the ships, *Kaladan* being a river in the Arakan region which flows into the Bay of Bengal at Akyab. However, from about 1948 several of the Henderson Line ships had been chartered to Elder Dempster who found them very suitable for their West African service.

97

Built by Lithgow's Ltd at Port Glasgow and propelled by a three-cylinder Rowan-Doxford oil engine, *Kaladan* was based on the very successful 10x10x10 Doxford Economy Ship design (i.e. 10,000 tons of cargo at 10 knots on 10 tons of fuel per day). Since being delivered to Paddy Henderson in October 1950 she had spent all of the six years on charter to Elder Dempster, trading between UK/European ports and West Africa. In late December 1956 *Kaladan* was homeward bound after some three months on the West African coast. The last loading port was to be Sherbro Island.

Sherbro Island lies off the south-western coast of Sierra Leone, some eighty miles south-east of Freetown, and consists mainly of mangrove swamps and jungle. It is separated from the African mainland by a stretch of water known locally as Sherbro River. This stretch of water is, in fact, a strait or sound which extends eastwards for about 30 miles, then swings south for 10 miles to meet the open Atlantic beyond the town of Bonthe, the main settlement on the island. The river, or strait, is navigable for vessels with a draught of up to 25 feet from its western entrance as far as the anchorage off Jamaica point and Bob's Island, a distance of some 25 miles. This is the usual anchorage for ocean-going vessels visiting Sherbro, cargo being sent out from Bonthe by lighter. While the river is 20 miles wide at its western entrance and about 5 miles wide at the anchorage, the navigable channel is quite narrow – not more than a quarter of a mile approaching the anchorage.

Bonthe was originally a 19th century British anti-slave trade post and was initially settled by freed slaves. It later grew into a centre exporting palm kernels, piassava and coffee. But problems of severe silting in the Sherbro River, and poor land communications caused by the extensive coastal swamps, resulted in its being progressively overtaken by Freetown as the major port on that part of the West African coast. However, the occasional Elder Demps-

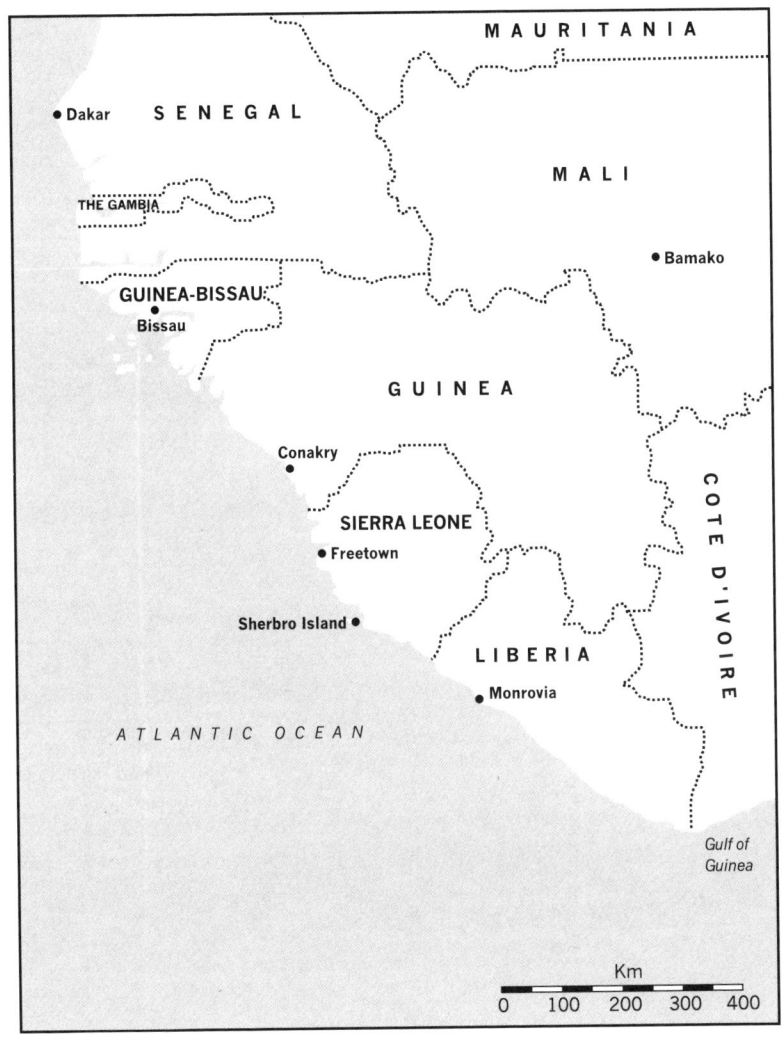

Map 7 West African coast and Sherbro Island

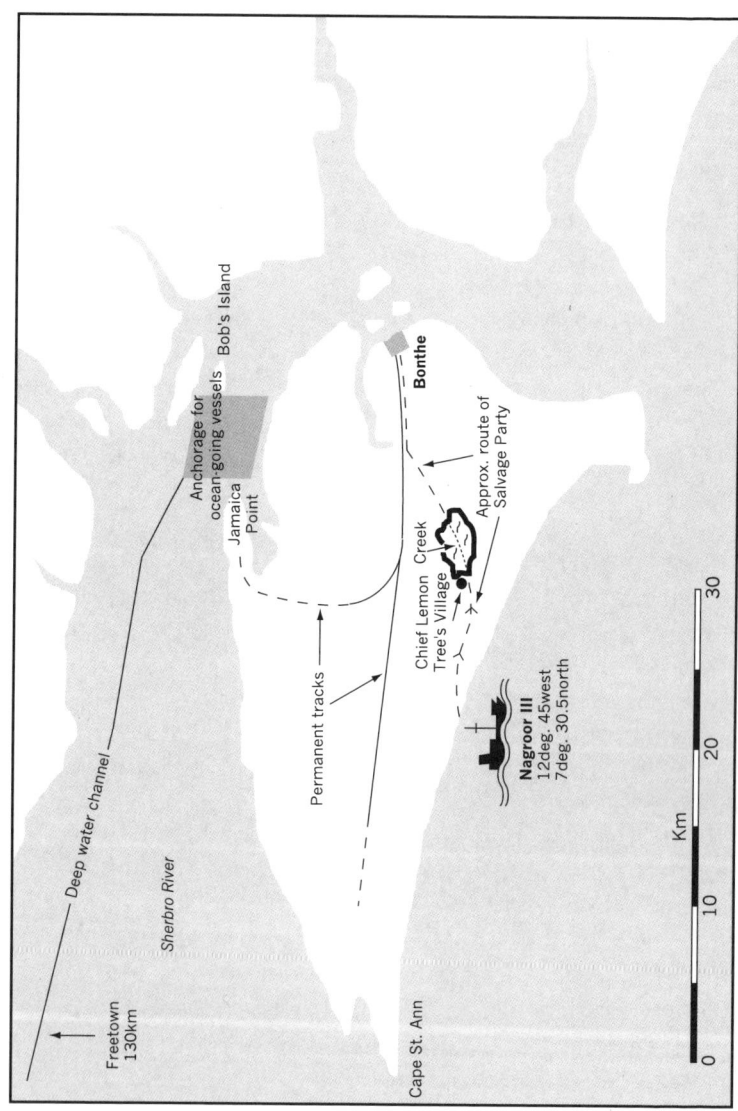

Map 8 Sherbro Island, Sierra Leone

Deep water channel

Freetown 130km

Sherbro River

Anchorage for ocean-going vessels

Bob's Island

Jamaica Point

Permanent tracks

Chief Lemon Tree's Village

Creek

Bonthe

Approx. route of Salvage Party

Nagroor III
12deg. 45west
7deg. 30.5north

Cape St. Ann

Km

0   10   20   30

100

ter and Palm Line ship, like *Kaladan*, still called in the 1950s to load piassava – a coarse fibre used in the manufacture of sweeping-brushes and brooms – and palm kernels.

After picking up our Sherbro River pilot at Freetown, *Kaladan* had steamed south to enter the river at its western entrance. Once in the narrow channel we sailed slowly along for about three or four hours, past mile after mile of mangroves, until we reached the anchorage between Jamaica point and Bob's Island. After the sort of delay that is not unusual on the West African coast, the first instalment of cargo from Bonthe eventually turned up in a couple of lighters. It proved to be a slow business working cargo at the Sherbro anchorage, mainly, I suppose, on account of the 20 mile round trip that the lighters had to do, so it took more than two days to load our 300 tons of cargo. While this topside cargo work was going on, the engineers were busy down below, fine-tuning the Doxford engine ready for the '10 knot dash' to the London River. At the end of our working day we would climb up out of the engine-room in our sweat-soaked overalls and, sitting on No. 3 hatch just forward of the galley, down a deliciously cold McEwan's Export. In the heavy, humid heat we gazed at the silent, almost menacing mangrove swamps and knew in our hearts that Sherbro Island was never going to become one of Thomas Cook's 'holidays with sun and fun' destinations.

But all that was in the past and I was brought back to the present by Clive poking his head round my door again and saying, 'Oh, by the way, there is a Nigerian cargo ship, *King Jaja*, being towed into Monrovia with some sort of engine damage. While you're down that way have a look at her and see what it's all about.' Monrovia is only some 150 miles from Sherbro Island so it appeared a fairly obvious and simple thing for me to attend both vessels. However, I should have remembered that nothing is simple or obvious in West Africa.

The only information on *King Jaja* was the following brief casualty report from Lloyd's agents at Monrovia:-

Monrovia 17 July

Motor Vessel **King Jaja** *(Amsterdam for Calabar) completely immobile with damaged thrust bearing off West African coast near Monrovia. Plan to use motor vessel* **River Benue** *on south bound voyage to tow* **King Jaja** *into Monrovia for repair.*

Lloyd's Agents.

Next day I presented myself at the Sierra Leone High Commission in London and was told by a frightfully bossy little clerk in the visa section that it would be quite impossible for them to issue me with an urgent visa in anything less than four days. Well, I did not have four days to waste while they moved my application round the office from desk to desk, so I decided to fly straight to Monrovia, where there were no visa requirements, and carry out an inspection of the damage on *King Jaja*. At the same time, I would try and get the local Lloyd's agents to use their influence and get me a visa for Sierra Leone. I had a hunch that they would probably get one the same day.

On arrival at Monrovia I was collected at the airport by the agent and taken down to the port. *King Jaja* had been towed in the day before and was berthed alongside the quay. As soon as I stepped on board I sensed something familiar about her and, on entering the engineers' accommodation, I immediately knew that I had been there before. Later, in the course of conversation, the Chief Engineer informed me, that before being taken over by her present owners, Nigerian National Line, she had been an Anchor Line ship.

'Yes,' I said. 'I know. She is the old *Tyria*. I joined her as

4th engineer in June 1955 when she was in the final stages of completion at Doxford's Yard in Sunderland. I stayed in her until I went ashore to sit for my 2nd engineer's ticket in August 1956.'

'Well, I never . . .' replied the chief. 'It's a small world indeed!'

As I gazed around the familiar old ship my mind was momentarily elsewhere. I was back on the delivery voyage from Sunderland to Glasgow, 'north about' through the Pentland Firth, then down through the Minch on a beautiful July Sunday afternoon. Over to port were the blue mountain-tops and ridges of Wester Ross, Torridon and the Cuillin; to starboard the misty line of the outer islands 'Wi heather honey taste upon each name'. Yes – even in hot and humid Monrovia I could, in my imagination, smell '. . . the tangle of the Isles'.

Forcing myself back to the business in hand, I asked the chief about his problem. He reported that they were on a loaded passage from Amsterdam to Calabar. Apparently, everything had been fine until just north of Monrovia on 17 July, when severe overheating was suddenly detected at the main-engine thrust-bearing. The engine was stopped and a subsequent inspection showed that the thrust-bearing was badly damaged. The master managed to anchor the vessel and the owners, on being advised of the situation, arranged for one of their other vessels, *River Benue*, to tow the disabled *King Jaja* into Monrovia.

Down in the engine-room I had a look at the problem. The main-engine thrust-bearing was a mess. The ahead thrust face-pads were all burned out and the thrust collar itself heavily scored owing to metal-to-metal contact with the burned-out pads. The damage had apparently been caused by a piece of broken glass, probably from a portable light, becoming lodged in the oil supply nozzle to the ahead thrust face, completely cutting off the oil supply. The little piece of glass had possibly been working its way through the system for some time. It was just unfortunate that it had ended up in a vital oil supply line.

The method of repair was to machine the damaged thrust collar by setting up a special grinding tool on the thrust block and then running the engine 'slow astern' while the grinding tool was moved back and forward across the damaged face until a satisfactory finish was achieved. A specialist firm in Glasgow, Nichol & Andrew, were engaged to carry out the repair and they had two of their technicians in Monrovia by the following day.

While I was busy with the burned-out thrust-bearing, the messenger boy in the Lloyd's agent's office popped round to the Sierra Leone Embassy in Monrovia in his lunch hour and, by waving a ten dollar note in the right direction, got me an instant visa.

With the *King Jaja* thrust-bearing problem in hand, I chartered a light aircraft at Monrovia and flew north-west along the coast towards Sherbro Island to try and locate the stranded trawler. About an hour after taking off we found *Nagroor III*. She was aground on the Atlantic side of the island, approximately 20 miles east of Cape St Ann. The casualty was in the surf-line more or less beam on to the beach, with a port list of two or three degrees. From the air the trawler appeared to be more or less intact, with the fishing nets and associated gear still in place on the outrigger booms. This seemed to indicate that looting had not yet occurred, despite the fact that at low water the casualty probably dried out at the fore end and would therefore be fairly accessible from the beach. I suspected that the local population would be more interested in the ship's fishing equipment than the machinery, navigation and accommodation fittings and the apparently undisturbed condition of the fishing gear suggested that the casualty had not yet been visited by the local Africans. On completion of our 500 feet aerial survey, we returned to Monrovia.

Next day, I attended at the owner's office to see what I could find out about the circumstances surrounding the stranding.

The information provided was that *Nagroor III* sailed from Monrovia on 8 July 1972 to carry out shrimp fishing off the Liberian/Sierra Leone coast. On 12 July, while fishing off Cape Mount, Liberia, a radio message advised that large shoals of shrimp had been located off Sherbro Island, some 100 miles north of Cape Mount. Fishing gear was stowed and secured and the vessel sailed north during the night of 12 to 13 July, arriving off Sherbro Island at first light.

While preparing the fishing gear at about 7.00a.m. it was discovered that both the fish hold and the engine-room were partly flooded. The source of the water ingress was traced to heavy leakage at the propeller-shaft stern-gland. By this stage the water level in the engine-room had reached the level of the main-engine crankshaft and the engine had to be stopped. This deprived the vessel of the use of the main bilge pump as this was driven by the main-engine. Unfortunately, the stand-by bilge pump was unserviceable, so the vessel was now without any means of controlling the ingress of water.

The owners, who were being kept advised of the situation, ordered two of their other vessels, *Nagroor IX* and *Nagroor IV*, to stand by and assist the disabled vessel. A spare stand-by bilge pump was transferred to *Nagroor III* but all attempts to get it working failed and the water level in the fish hold and engine-room continued to rise. At or about 4.00p.m. the crew were taken off by *Nagroor IX* and, about one hour later, the casualty stranded in the surf on the south-west side of Sherbro Island.

Based on the aerial survey and the information provided

by the owners at Monrovia, which suggested that the hull was intact apart from leakage at the stern-gland, I reckoned that salvage and repair of the vessel should be possible within the insured value. The outcome of our meeting at Monrovia was that the owners entered into a contract with Bureau Wijsmuller Towage & Salvage Co., who had a salvage tug on station at Cape Verde. The contract terms were that the tug would proceed to the casualty location via Freetown at a cost of £1,000 per day, station to station. Should salvage prospects prove to be favourable, the daily rate contract would be converted to a Lloyd's Open Form, No Cure – No Pay Contract.

From Monrovia I flew to Freetown, to await the arrival of the salvage tug. The Sierra Leone authorities proved to be about as co-operative with regard to granting clearance for the tug to operate in their waters as their High Commission in London had been about issuing me with a visa. Clearance for the tug was only obtained after two days of fairly high-level activity and the intervention of the Naval Attaché from the British High Commission. During the two days spent at Freetown I met the owners' representative, a Panamanian gentleman who rejoiced in the unlikely name of Eddie MacTaggart. Another important player who arrived on the scene at this stage was Leen Breukel, the Bureau Wijsmuller salvage officer.

MacTaggart and I spent a couple of days in the same Freetown hotel and during that period I noted two points of interest. Firstly, Eddie was a confirmed teetotaller, and secondly, he was quite unable to tie a knot in a necktie. In the marine business the former point of interest was probably the more unusual of the two. The hotel was not particularly pretentious but they did insist that a long-sleeved shirt and tie (Planter's Rig) be worn in the dining-room at dinner. On his first day in the hotel, and wearing a rather loud short-sleeved shirt, Eddie was stopped and asked to please wear a tie. Unfortunately, at that time, he did not posses such an exotic item of attire and had to borrow a clip-on tie from one of the hotel staff. Next morning, he sallied forth into the Freetown shopping area intending to purchase a tie. This he managed to do but, unfortunately, was unable to obtain one of the ready-knotted clip-on jobs and had to make do with a traditional 'tie-it-yourself' model. This, however, was rather a problem as he had never owned such a thing, and try as he might, he was quite unable to tie a knot in the confounded thing. This rather embarrassing state of affairs he confided to me over a cup of tea on the first afternoon of our meeting. As a result, before dinner that evening, I popped into his room to show him how to 'tie' his tie.

This operation proved not to be quite as simple as I had thought. I suppose it must have been because I was looking at the tie from the outside rather than from the inside, as it were. We only managed to solve the problem by assembling the tie round my neck with a very loose knot and then transferring it to Eddie's neck so that all he had to do was pull it up tight. It was, as I reflected later, not the sort of problem an underwriter's surveyor meets every day.

The salvage tug *Jacob Van Heemskerk* arrived at Freetown from Cape Verde on the evening of 25 July. Breukel, MacTaggart and I boarded shortly after arrival and we sailed for the casualty location around midnight. We spotted *Nagroor III* at about 10.00a.m. on the following morning and by 11.00a.m. had anchored in some five fathoms of water about 600 metres from the casualty. While steaming along the coast after daybreak, both the salvage officer and I had been rather disturbed by the violence of the surf breaking on the beach. However, where the casualty was stranded the surf did not appear to be quite so strong. At least, we optimistically convinced ourselves that this was the case.

The plan of action was that Breukel, MacTaggart and I would proceed to the beach in a Gemini, which is an inflatable craft capable of carrying five or six persons and powered by an outboard motor. Having reached the beach we would then board the stranded vessel; it was almost low water and access from the landward side would not be much of a problem. We were obviously going to get a bit wet during the ride to and from the beach and while inspecting the casualty. I therefore decided, in my infinite wisdom, to wear only my shorts and a pair of plimsolls – the day of the designer trainer had not yet dawned – the latter being to protect my feet while clambering around the casualty. In addition, we all had lifejackets, which turned out to be a wise precaution. However, my decision to leave my long-sleeved bush shirt on the tug was to prove unfortunate to say the least.

We boarded the Gemini at about noon and headed off towards the golden sand of the beach. The Gemini was manned by the tug's chief officer, Cees Blaauw, and the 2nd engineer, Willem Fossen, making a total of five in our party. The Gemini rose and swooped over the big Atlantic rollers as we hurtled towards the beach. From this quite low-level position the surf did not appear quite so benign as it had looked from the solid deck of *Jacob Van Heemskerk* – in fact, it now looked positively ferocious. I sat in the bottom of the Gemini with my back against the starboard side, my right arm along the top of the side and with my hand firmly grasping the lifeline looped along the outside of the craft. I was looking forward towards the beach, watching the palm trees grow larger as we approached, when someone said, 'Here's a big one'.

I glanced aft and saw, towering above our frail little craft, a wave of massive proportions. I looked back towards the beach, wondering how long it would take us to get there when, suddenly, everything went crazy. I was dimly aware of being catapulted forward and turning over and over, almost in slow motion, in the midst of a grey mist. I was also aware that my right hand was entangled in the Gemini lifeline which I had been holding tightly before the wave struck. I felt my head bump against the bottom boards of the inverted Gemini a couple of times. Suddenly, my hand came free and I shot to the surface some 50 metres from the beach. I immediately struck out for the shore with the desperation of a man pursued by the entire aquatic cast of *Jaws I* and *II*! Then I spotted Leen Breukel a couple of metres away – standing up! I stopped my frantic thrashing about, put my feet down and also stood up – to be promptly knocked down again by the next incoming breaker. Breukel and I eventually stumbled up the beach to discover our three companions emerging from the surf at about the same time. The inflatable was washing back and forth at the water's edge so we all grabbed it and pulled it ashore. Then we sat down and in complete silence, except for the pounding of the surf, matched only by the pounding of our hearts, surveyed the situation. It was not good; it was not good at all.

After some time, MacTaggart, Breukel and I decided to walk along the beach to the reason for our journey, which lay some 200 metres away, and carry out our inspection. Meanwhile, Blaauw and Fossen started a careful inspection of the Gemini to check what damage had been sustained when it capsized.

The three of us were able to scramble on board without much difficulty but, once there, were dismayed to find that extensive looting had taken place. All the navigation and radio equipment had been torn out and removed. A similar state of affairs was found in the galley and accommodation spaces. Only the steel-casing bulkheads remained. All copper fittings in the reefer spaces and the reefer machinery room had been torn off and removed. The engine-room was flooded to sea level, presumably through the defective propeller shaft stern-gland. At this stage the trawler had been stranded for about ten days and the local natives had done a thorough looting job in that time. Strangely, though, they had ignored the fishing gear; perhaps they intended coming back for it. But where were they now? The beach, apart from our two companions crouched round the Gemini, was deserted, and the silent, green jungle along the edge of the sand showed no signs of life.

Having completed our survey we assessed the prospects of salvage and repair. The condition of the casualty, its position on the beach and the fairly remote geographical location of Sherbro Island, all persuaded us that economical salvage and repair would not be possible. To say that this was a disappointing conclusion to reach, after all the trouble we had taken to reach the trawler, would be an understatement. We then scrambled back over the port side of the trawler, dropped down onto the sand and headed back along the beach to rejoin the others at the Gemini. The situation there was not very optimistic. The outboard motor had been submerged at full power and was completely seized. In addition, the ignition system was totally saturated in salt water. Cees Blaauw, the mate, had brought a portable two-way radio from the tug so that we would have radio contact with them, but that had been lost when we capsized in the surf. Without the radio we were unable to communicate our situation to the tug and without the outboard motor it was going to be difficult to get back. There were two paddles secured inside the Gemini but they would not be enough to get us through the surf. While the three of us had been inspecting the casualty, one of the others at the Gemini had spotted a native watching us from further along the beach. On the off-chance that there might be some habitation in that direction, Willem Fossen and I set off to see what we could find. We were anxious to try and obtain some extra paddles, or at least some timber from which we could fashion some makeshift ones. Our luck was in. After scouting along the beach for about a kilometre we came across a deserted hut with a dug-out canoe lashed down outside it. Sticking out of the sand beside the canoe were two splendid paddles. There was no sign of the owner, so we confiscated them for the common good and hurried back along the beach to rejoin the others.

By now we were feeling the effects of the hot African sun and the brisk onshore wind. I was beginning to regret having left my shirt on my bunk in the tug. Fortunately, the life-jackets gave some protection. The others were reasonably clad. Willem Fossen was wearing his tropical lightweight overalls, but he had lost one of his shoes when we capsized. Breukel was wearing his working uniform of shirt and long pants. On his feet he had leather sandals of the type favoured by devotees of Hare Krishna. They were to prove insufficiently robust. MacTaggart was attired in T-shirt, long pants and, under his shirt, a string vest.

We rested on the sand and studied the behaviour of the surf. Our problem was how to get the Gemini and, equally importantly, ourselves, through the pounding waves and back to the tug with only four paddles as our means of propulsion. After some time, we noticed that there was a sort of pattern to the breakers as they swept in from the Atlantic. There appeared to be four more or less equally spaced waves, then a bit of a longer gap before the fifth one flung itself on the beach. We therefore concluded, somewhat optimistically as it turned out, that if we picked the right moment there would be time to paddle clear of the surf in the gap between the breakers.

The wrecked outboard motor was made secure and the inflatable dragged down into the water. We pushed it out until we were about knee deep; then, when we judged our timing was right, leapt aboard and paddled like men possessed. Almost immediately, we were struck by a huge wave and thrown violently back into the surf.

We collected ourselves back on the beach, righted the Gemini, recovered our four paddles – which had been sent flying in all directions – then paused to consider our next move. Someone said, a bit superfluously, 'I think we got our timing wrong'. After studying the surf pattern again we decided to try once more. This time we pushed the Gemini out a bit further while we waited for the right moment. The end result, however, was just the same – an explosion of foaming water with bodies and paddles flying in all directions. Twice more we tried, but with the same lack of success. On the final attempt my life-jacket was torn off. We were now becoming exhausted and both Breukel and I decided that any further attempts would be useless and could result in one or more of our party being seriously injured.

Our activities had not gone unnoticed on the tug. They had been following our progress fairly closely since we had left them at noon. During our attempts to get through the surf they had put their lifeboat in the water; it was now standing off the beach, just outside the surf. By means of hand signals we notified the lifeboat that we required some food and water and would be spending the night on the beach. A rocket line was then sent ashore, followed by a watertight drum containing food, water and a signalling torch so we could communicate with the tug after dark.

Only an hour of daylight was left so we quickly set about making camp in as sheltered a spot as we could find among the sand dunes. We scouted up and down the shoreline and soon found enough wood to get a good fire going. As the sun went down the wind, which was still fairly brisk, made us feel quite cold after the heat of the West African day. Breukel, who had recently completed a survival course in Holland, had us all digging bivouac holes in the sand and building wind-breaks from bits of palm-trees. Despite the protection of my bivvie-hole and wind-break I felt quite cold without my shirt. However, Eddie MacTaggart rose to the occasion and, after whipping off his T-shirt, handed me his string vest – greater love hath no man than he give his string vest to another. Yes, he wasn't a bad sort of lad, was Eddie.

Fortified by a meal and the comforting effects of our fire, we discussed the plan for the morning. Departure through the surf appeared to be a non-starter so the only alternative was to walk through the bush to Bonthe. We estimated the town to be about 25 miles away – 30 at the most. But how to get there through 25 miles or so of jungle and mangrove swamps? Whilst collecting wood for our fire we had spotted a couple of natives in the direction of the fisherman's hut where we had borrowed the paddles. Our plan was to try and make contact with them first thing in the morning and see if they could guide us to Bonthe.

Having thus decided, Breukel signalled our intentions to the tug. They immediately acknowledged our message as being received and understood. We then settled down to get some sleep, after what had been a rather eventful day. This, however, proved difficult owing to the combination of painful sunburn and squadrons of mosquitoes which appeared on the scene just after dark. We whiled away the hours collecting wood to keep our fires going. The moon was up, making it relatively easy to move about. My gaze frequently shifted to the twinkling, riding lights of the anchored tug. I could imagine the captain and the chief engineer sitting comfortably, discussing our predicament over a glass of Heineken or, even better, a single malt. The old phrase 'so near and yet so far', sprang to mind.

Between our firewood collecting forays, Eddie Mac-Taggart entertained us with tales of an earlier career with the Panama City Police Department (PCPD). The standard issue personal weapon was the 38 calibre, Smith & Wesson revolver. 'Just a goddamned pea-shooter,' opined the good Eddie. 'I tell you guys, I allus packed a Colt 45! Now *that's* a weapon – hit some jerk with that sucker and he goes down and stays down!'

While with the PCPD he had spent some time with the vice squad. When raiding illegal gambling establishments the officers would rush in yelling, 'Raid! Raid! Raid!' – followed by Eddie, firing his Colt 45 in the air.

'Man, it sounded like a goddamn cannon,' he said with satisfaction. 'Scared the crap right out of those SOBs and they were out the door like greased lightning. Left their dollar bills laying right there on the table too.'

'So what happened to all the money?' I asked. Eddie gave me a pitying look.

I finally fell asleep about 4.00a.m. only to be interrupted by Leen Breukel shaking me awake about 6.00a.m., as the first grey light of dawn lightened the eastern sky. After a bite to eat and some orange juice, I set off for the fisherman's hut. A thin wisp of smoke was rising from near the hut and, sure enough, there was our elusive African getting ready to have his breakfast, which looked a lot more appetising than ours. He was quite friendly and obviously intrigued by our presence. No doubt he had observed our adventures of the previous day and was well aware that we would have to get to Bonthe. We had a major language problem but no great difficulty in indicating to him what we wanted to do. In the best Robinson Crusoe style I drew a map of Sherbro on the sand, showing our position and the estimated location of Bonthe. By sign language I communicated our wish to proceed there and asked if he would take us. He nodded his head vigorously as he emitted a stream of incomprehensible words. I took it all to mean, 'No problem, man'.

The watertight ration container which the tug had sent

ashore was quite impossible to carry through the jungle, so we distributed the remaining contents amongst ourselves. My portion was a long loaf of bread in a cellophane bag. Having emptied the drum we presented it to our guide in the form of payment on account, together with a spare lifejacket which we had somehow acquired.

We were ready to leave the beach at about 7.00a.m. The tug was still at anchor, no doubt watching us and trying to guess what we were up to. The 2nd engineer's missing shoe suddenly became a serious problem now that we were about to commence a 25 or 30 mile trek through the jungle. He maintained he would be fine and we moved off. Almost immediately, we found on the sand the sole of an old rubber flip-flop. He whipped off his sock, put the rubber sole in position against *his* sole and put his sock back on over it. Willem did a test walk up and down and gave us the thumbs up! With a glance back at the tug, we followed our little guide off the beach and into the jungle.

At first we walked along a fairly distinct track but after about one kilometre this vanished into a swamp. Our guide did not seem too bothered by this; he plodded steadily on and we followed in his wake. All morning we moved resolutely through the swamps. Occasionally we emerged onto a dry section of ground but it was not long before we were back in the swamps. Sometimes the water was knee high, sometimes waist high. To the right and left of us monkeys chattered and birds called out – sometimes raucously, sometimes musically – as they wondered who these strange, alien invaders were.

Now and again we stopped for a drink. I was still carrying my loaf of bread – most of the time on my head – as we navigated the mangrove swamps. The 2nd engineer's temporary rig on his left foot was standing up quite well but the mate's leather sandals were showing signs of strain.

About five hours after leaving the beach we suddenly

came upon a small African village and were quickly surrounded by a curious crowd. The guide showed no signs of proceeding on our journey and, for some time, we could not understand why he had brought us into the village. After standing around in confusion, surrounded by excited villagers, we were slightly unnerved by the approach of a most unusual looking character. He was wearing a Hawaiian-type shirt, a pair of baggy shorts and, on his head, a blue-and-white striped pom-pom hat. There was a strange, but faintly familiar smell about him; I couldn't quite place it. He spoke a few words of English and was thus able to advise us that he was Chief Lemon Tree! He seemed to have some difficulty in understanding that we wanted to get on the road to Bonthe without delay, and kept inviting us into his hut for a bite to eat and something to drink. Eddie Mac-Taggart was not impressed with this invitation from one of his ethnic brothers.

'Hell! We don't have time for that, Chief,' he said. 'We got to get to Bonthe.' And in an aside to me, 'We'll be in the cooking pot for sure if we don't get out of this goddamned place.'

Just about then I noticed the Chief's second-in-command, who was a particularly villainous looking individual, giving the mate, Cees Blaauw, a gentle poke with his finger. The mate was a fairly chubby chap and would probably have provided the tastiest dish. As he poked, he smiled and, in so doing, displayed a ferocious arsenal of filed teeth. MacTaggart and I looked at one another like two horrified Jack Bennys; MacTaggart's normally smiling, ebony features were now a decidedly unhappy shade of battleship grey.

After much to-ing and fro-ing, mostly around Chief Lemon Tree's hut, we became more and more uneasy about the villagers' intentions. At the back of my mind was the growing suspicion that these were probably the people who

116

had looted the trawler and I hoped they were not thinking 'dead men tell no tales'.

Then, like a light bulb coming on, the reason for our diversion into the village became crystal clear. I saw we were on the edge of a large creek, which would have to be crossed for us to continue towards Bonthe – and for that we would require the use of one of the village canoes. This could only be provided with the consent of Chief Lemon Tree. A canoe was produced and pushed down to the edge of the water, but – wait a minute – a last snag! The Chief wanted to be paid, both for our use of the canoe and for the continued services of the guide who was, presumably, one of his subjects. This posed a real problem because, among the five of us, we did not have a single cent. As we pondered our next move, MacTaggart rose to the occasion.

He had recognised, before the rest of us did, that the cause of the Chief's confusion was the over-generous consumption of palm toddy, which also explained the somewhat familiar aroma that enveloped him.

'Now don't you worry your ole head about that, Chief,' said MacTaggart. 'We'll send the money back from Bonthe with the little fellah here,' indicating the guide, 'and a bottle of drink for yourself.'

Lemon Tree beamed at the mention of drink – here was a man who understood his needs – but, in case there was any confusion, he would like a contract in writing. I had been carefully carrying the loaf of bread for the last five hours and in the plastic bag was a buff-coloured slip of paper with, presumably, the packer's number on it. All I needed was a pencil. From the depths of one of the huts a little stub of a pencil was produced and handed to me. It is now nearly thirty years since I stood in that African village but I can remember quite clearly what I wrote on that slip of paper. Leaning on the square kapok of Eddie MacTaggart's lifejacket, I wrote:-

117

To Chief Lemon Tree

Will pay one pound for the services of your guide and two pounds for the use of your canoe. Will send back from Bonthe one bottle of drink for yourself.

Signed S. Walker
Lloyd's of London
27 July 1972

Chief Lemon Tree received the 'promissory note' with all the gravity of a London underwriter. After peering at it for several moments, he beamed once more and gave orders for the canoe to be pushed into the water. The five of us heaved a collective sigh of relief and prepared to board our vessel. To further seal our contract and, as I was fed up with carrying it all over West Africa, I solemnly presented the loaf of bread to the Chief. We then pushed off into the creek and, as we looked back at the smiling, waving villagers, MacTaggart whispered to me, 'I thought we was all for the cooking pot back there, I surely did.'

We must have spent about an hour in the canoe as the boatmen threaded their way through a complicated maze of mangroves before eventually grounding close to a path which, we were given to understand, eventually led to Bonthe. Sign language indicated that we would reach our destination when the sun was low in the sky. Whether that would be today or tomorrow was not clear.

The going was much easier on this stage of our journey. We had left the swamps behind and the path through the forest was quite reasonable. The temporary rig in place of Willem Fossen's shoe was holding out but Cees Blaauw was having trouble with his leather sandals, which were slowly falling apart. MacTaggart had hurt his leg somewhere along the trail and was limping quite badly. All in all, we must

Typical scene in the Rangoon River with Asiatic Steam Navigation Company's *Bahadur* working cargo, June 1958.

Typical Burmese country craft at Bassein, March 1958.

The author at the Yala Game reserve lodge

Universal Trader wrecked on Little Basses Reef, March 1968

Universal Trader showing where the hull broke in two at the bridge front and the engine room forward bulkhead

Various views of *Kostis A Georgilis* aground and on fire at Great Coco Island, November 1967

Gero Michalos sunk at Akyab, July 1968

Fore deck of *Gero Michalos* with residents of Akyab and district waiting for the tide to drop a bit further and allow them access to the rice cargo. The forecastle deck including the windlass has been completely stripped

A West African dug-out canoe similar to the one provided to the salvage party by Chief Lemon Tree

Last photograph of *Evdokia* taken during the crew evacuation at about noon on 11 June 1979

Photo courtesy Natal Mercury

The upside down mid-ship house and bow section amid the foaming surf and rocks of South Africa's wild Tsitsikamma coast

Photo courtesy Natal Mercury

Causeway Adventurer arrives on 1 May 1980 and work commences making towing connection fast to stern bollards

The stranded *O Yang 77* with the settlement of Waitangi in the background

Ox carts at Bagan

The dry season low river level has exposed sand banks on which a temporary village has been built with an itinerant population engaged in providing sand to the building industry to make cement

The temple-studded Sagaing hills on the Irrawaddy opposite Mandalay

have presented a sorry sight: more like a motley collection of beggars than a marine salvage party.

Since leaving the beach we had been moving in a general north-easterly direction. Then, after a couple of hours of trudging along the path from the creek, we reached a main east/west track. Our little guide turned right and headed off eastwards, indicating that Bonthe lay somewhere ahead. After a little while we began to see native huts here and there along the sides of the track, and then we met a sophisticated young man who spoke excellent English. He confirmed that we were on the right road to Bonthe; it was, he said, only four or five miles away. Our original plan was to call at the district officer's house and see if we could contact Freetown. I had hoped that *Jacob Van Heemskerk* might have steamed round from the casualty location and be waiting for us at the Bob's Island anchorage in the Sherbro River. Based on his knowledge of the captain, Cees Blaauw thought that this was unlikely. He was correct – the tug had gone to Freetown.

Our plans were altered somewhat when our new young friend told us that on our way into Bonthe we would pass a big mission.

'What kind of mission?' I asked.

'Oh, a Roman Catholic mission, run by a white man.'

On receipt of this interesting piece of information, Willem Fossen and I pushed on ahead, leaving Breukel to look after the limping MacTaggart and Blaauw. In about an hour we were in sight of the town and there, just off the track, were substantial stone-built buildings and a church. Willem and I approached what appeared to be the main building and knocked on the door. After a few moments it opened and a smallish, European man looked out. He was wearing a short-sleeved, open-necked shirt and khaki shorts. He peered curiously over the top of the half-moon spectacles that were perched on the tip of his nose.

119

'Good afternoon,' I said. 'My friends and I – there are five of us altogether – have walked from the Atlantic side of the island where we were marooned when our boat capsized yesterday. We are a little bit weary and wondered if we could have a rest and perhaps a cup of tea.'

He appeared entirely unfazed by our unheralded arrival and startling appearance.

'Come away in,' he said, and led the way into the house. We were shown to a table where our new acquaintance had obviously been reading a copy of the *Irish Farmers Weekly* and drinking a Heineken lager. Gratefully, Willem and I sat down and waited for the others to arrive. Father Curran, for that was the name of our host, sent for his house-boy, Joseph, to make us some tea. One by one our three companions, together with our faithful guide, arrived and collapsed into the available chairs. For a while we sat in silence, occasionally heaving the odd sigh of relief. I took off my plimsolls and saw the reason for my increasing discomfort around my feet. Hours of walking through the swamps had filled the shoes with sand which had gradually rubbed the skin off the knuckles of all my toes. It was heaven to take the shoes off, but I was unable to get them back on.

The mission included a medical section and, although they were away doing their rounds of the villages, Father Curran and his assistant, Father Burns, soon tended to our various lacerations, abrasions and patches of acute sunburn. Father Burns was then instructed to prepare a meal for us. It turned out to be sausage, bacon and eggs for five! Despite the fact that it was swimming in grease, it was undoubtedly one of the finest meals I have ever eaten.

While we were eating we explained our situation to Father Curran. He had provided me with one of his clean, white shirts so I was able to divest myself of the lifejacket and MacTaggart's string vest and look more or less respect-

120

able. The mate and 2nd engineer had also been more suitably shod.

In Bonthe itself, there was a bungalow that served as a mess for a mining company during part of the year. It was, at that moment, unoccupied and, as it had permanent servants and ample accommodation for us all, Father Curran arranged for us to stay there until we could fly out to Freetown.

Before moving to the mess we explained that we had incurred some debts along the way. Would Father Curran be able to pay the guide and purchase a bottle of suitable drink for Chief Lemon Tree – on the understanding that the Lloyd's agent at Freetown would reimburse him for all expenses incurred on our behalf? No problem. Joseph was despatched to buy a bottle of Diamond Gin, apparently a favourite tipple with the rural locals.

Our stomachs filled, our hurts soothed and our debts dealt with, we spent an excellent evening on the verandah of the mining company mess, looking out over the Sherbro River and sipping our cold beer. All in all, it had turned out to be 'one helluva' day!

Next morning our guide said 'farewell' and with three Sierra Leone pounds and the bottle of gin, set off back to Chief Lemon Tree's village. Father Curran and I went round to the Police Post where we were able to contact Freetown by radio and advise the Lloyd's agent of our location. There is an airstrip outside Bonthe but only one scheduled passenger flight per week, and we had just missed it! However, arrangements were made for a passing aircraft to divert to Bonthe later that day to take us to Freetown.

About three o'clock in the afternoon Father Curran drove us out to the airstrip where a Fokker Friendship of the local airline whisked us to Freetown. I disembarked, still wearing my shorts and Father Curran's shirt and with my plimsolls hanging around my neck. Despite the ointment applied back at the mission the previous day, I still could not get my shoes on.

Before boarding the aircraft, Breukel and I agreed that a fair and reasonable payment to Father Curran for our board and lodging, medical treatment and other sundry expenses such as Lemon Tree's bottle of Diamond Gin would be £100. I told the priest that I would instruct the Lloyd's agent in Freetown to send back the shirt, washed and ironed and a cheque for £100. He protested vigorously, 'Oh, that's far too much, far too much!'

'Now, don't argue,' I answered. 'It's not every day a Roman Catholic mission gets one hundred pounds from a Presbyterian.'

He was still roaring with laughter as we trooped on to the plane.

From the airport outside Freetown, we proceeded to the harbour where *Jacob Van Heemskerk* was anchored. The mate, Cees Blaauw, and the 2nd engineer, Willem Fossen, returned to their everyday duties while Leen Breukel, Edwin MacTaggart and I recovered our kit, said goodbye to our two Dutch comrades and returned ashore to our hotel.

Next day, the tug sailed for her salvage station at Cape Verde, Breukel left by air for Amsterdam and MacTaggart for Monrovia. On my own and feeling suddenly rather lonely, I strolled along to the Lloyd's agent's office where, over a cup of tea, I gave him a run-down on the adventures of the previous few days. I handed over Father Curran's shirt, freshly laundered by the hotel staff, with instructions for it to be sent back to the mission, together with a cheque for £100.

For the benefit of anyone who might have been concerned at such lavish expenditure (London Office Accounts Dept), the following breakdown was provided:-

NAGROOR III   Case No 176/72

Survey Expenses incurred at Sherbro Island

Services of Guide provided by Chief Lemon Tree	£1.00
Use of canoe provided by Chief Lemon Tree	£2.00
Medicinal refreshment for Chief Lemon Tree	£0.50
Board and Lodging at Bonthe including supply of clothing and footwear	£96.50
TOTAL	£100.00

Once all that was out of the way I prepared a telex message for London, detailing the position and condition of the casualty and explaining all the reasons why salvage and repair would exceed the insured value. Apart from issuing my formal Survey Report, that, to all intents and purposes, was really the end of my involvement with *Nagroor III*.

I then flew back down to Monrovia to check on the status of the *King Jaja* repair. In contrast to the difficulties experienced up at Sherbro Island, everything had gone smoothly on my old ship. The repair was completed on 27 July and, after a satisfactory engine trial, *King Jaja* resumed her interrupted passage to Calabar.

Next day, I left Monrovia and returned to Southampton. Back in the office, I was involved with several routine surveys in the Southampton/Portsmouth area. One of them involved the flamboyant Lady Docker's old yacht *Shemara* which had run aground somewhere in the Caribbean and was under repair at Vosper Thorneycroft's yard.

In the days following my return from Freetown, I began to notice that every morning I felt decidedly under the weather, almost as if an attack of flu was in the offing: shivering, aching joints, etc. By afternoon these symptoms had worn off and I felt quite normal. Then, one morning, while discussing the *Shemara* repairs with the captain, I began to feel quite ill. Instead of easing off as the day went on, the shivering and aches and pains were becoming worse. I took myself off home to bed, advising the office that I definitely had the flu.

After twenty-four hours of alternately perspiring and freezing, the local GP was sent for. He took one look at me and dispatched me by ambulance to Winchester Hospital where they very quickly diagnosed malaria. I was an object of considerable interest to the medical staff as the incidence of malaria in the wilds of Hampshire is not frequent! After about a week of treatment I was discharged, feeling very weak and wobbly indeed.

Sherbro Island had finally finished with me.

6

In Northern Waters

I spent all of 1973 attached to the Salvage Association Southampton office but actually spending nearly all my time travelling to inspect damaged ships in various parts of the world. In January, I was at Quintero Bay in Chile to look at a British tanker with a fire-damaged engine-room. One month later, I was at Luanda to check the condition of a Greek cargo ship with a main-engine scavenge fire problem as well as a case of spontaneous combustion in her cargo of palm kernels. That was followed by another spontaneous combustion cargo problem in a Yugoslavian ship at Mauritius.

Back in the office at Southampton I was busy completing my report on the Yugoslavian ship when the phone rang. It was the chief surveyor in London wondering if I would like to spend a month in the Glasgow office where they were a bit short-handed owing to sickness. It was years since I had spent any time in Glasgow – not since I had gone off to Singapore to join Ritchie & Bisset in 1961 – so I said 'yes'.

Life in the Glasgow office proved to be fairly quiet, a few routine surveys in connection with Shipbuilders' Liability Policy claims at Yarrow's, Lithgow's and the Govan yards kept me occupied on a 9 to 5 basis. Then, on 5 May, the livestock carrier *St Rognvald*, operated by North of Scotland Orkney & Shetland Shipping Co., Ltd., ran aground at

Kirkwall in the Orkney Islands. Bill Blackburn, the principal surveyor of the Glasgow office, went up to Kirkwall the day after the stranding and made an initial assessment of the situation. On returning to Glasgow he told me that I would be proceeding to Orkney as soon as the salvage operation commenced.

Bill was one of the Salvage Association's 'characters'. He was a man of strongly held opinions, which he frequently expressed in a most forthright manner. All of us, from the most junior typist right up to the very chairman himself, were liable to be on the receiving end of his, frequently unconventional, views. On this particular occasion Bill, or 'Blackie' as he was known in the trade, was very concerned that the salvage operation on *St Rognvald* should proceed smoothly. Looking at me from behind his large and imposing desk he issued a stern warning. 'There have been too many foul-ups with salvage cases at this office recently, Mr Walker. I don't want any more.'

One of these 'foul-ups' had concerned the puffer *Raylight*. The Clyde puffer, immortalised in the 'Para Handy' stories by Neil Munro, was a sturdy steam-coaster, approximately 70 ft in length, which could carry 120 or so tons of cargo and deliver it with her own gear wherever there was enough water to float. If there was not enough water she could beach herself, unload the cargo and sail off on the next tide. The very first of these vessels had non-condensing engines, the exhaust steam being ejected from the funnel; the resulting sound caused them to be termed 'puffers'. Despite the switch to surface condensing engines, and later to diesel-power, with the consequent elimination of the 'puff-puff' noise from the funnel, the name 'puffer' has remained.

Raylight, a fairly modern diesel-driven puffer, had suffered an engine breakdown in gale force weather in the Sound of Jura on Christmas Day 1972. The master managed to anchor close to Jura's east coast, but the south-east gale caused her to drag and ground at the entrance to Tarbert Bay. The rudder was lost, the propeller damaged and the hold and engine-room flooded as a result of damage to the bottom plating. After pumping out the flooded spaces the casualty was refloated on 30 December and towed to Crinan by her sistership, *Dawnlight*. The intention was to moor the casualty alongside the pier at Crinan until a tug became available to tow her round to one of the Clyde yards for repair. However, in view of the Scottish hogmanay celebrations the tug was not due to arrive until 3 January.

What happened after *Raylight* arrived at Crinan is best described in the marine casualty reports published in Lloyd's List of 2 January 1973.

Glasgow 30 December

Motor vessel Raylight *refloated PM today and towed to Crinan where now lying safely afloat under easy controlled pumps. No tug available until Wednesday (3 Jan) therefore* Raylight *remaining constantly manned until then.*

Salvage Association Surveyors

Glasgow 01 January

Motor vessel Raylight *sunk alongside pier at Crinan. Vessel now sitting on the bottom with 45deg. list to port. Attempts to pump her dry will be made at low water. No oil pollution.*

Coastguard.

If my memory serves me right, I think it was Jack Donnelly who was the salvage surveyor on the job. Over at Jura he had done an excellent job in what were fairly atrocious conditions, rigging salvage pumps in the hold and engine-room and generally supervising the refloating of the puffer and the subsequent towage to Crinan. Once berthed at Crinan, he issued all the necessary instructions to the crew to enable *Raylight* to lie safely afloat until the tug arrived on 3 January. Everything seemed to be under control. The salvage pumps in the hold and engine-room were coping quite easily with the leakage from the damaged bottom plating. Bearing in mind that it was the great Scottish festival of hogmanay, Bill Blackburn felt understandably justified in telling Jack that he should go home for a couple of days, then return to Crinan on the 2nd to make sure that everything was in order for the tow round to the Clyde. After all, what could possibly go wrong in that safe little West Highland harbour. Those of us who have read and enjoyed Neil Munro's masterpiece, *In Highland Harbours With Para Handy* can appreciate that in the world of West Coast puffers quite a lot can go wrong at any time of the year – never mind hogmanay!

I could just imagine the skipper, mate and engineer convincing themselves on hogmanay that everything was fine and dandy, then slipping up to the Crinan hotel to bring in the New Year in the customary style. I have a picture of the three of them weaving their way back down to the pier in the early hours – or, more likely, the not so early hours – of New Year's day, to find only the wheelhouse and funnel showing above the water. Their horror and total disbelief can best be imagined. I will bet they never tasted another drop for the rest of the year. Well, at least not for the rest of the week anyway.

The second 'foul up' concerned the Caledonian Mac-Brayne inter-island ferry *Loch Seaforth*. While on passage from Lochboisdale in South Uist to Tiree, this vessel stranded on rocks in the Sound of Gunna, which separates the islands of Tiree and Coll, in the early hours of 22 March

1973. The bottom was holed and, as a result, the engine-room flooded. The tug *Cruiser* answered the distress call and arrived at the scene shortly after first light and by 10.00a.m. had, with the assistance of the rising tide, successfully towed the casualty clear of the rocks. With the engine-room still flooded *Loch Seaforth* was berthed alongside the pier at Scarinish, where the depth at high water was 18 feet. The tug *Cruiser* had no salvage pump onboard and was unable to assist with de-watering the flooded engine-room. It was therefore released to return to its normal business which, of course, had been interrupted when it answered the 'Mayday' call from *Loch Seaforth*. At the same time the tug *Warrior*, with a 50 ton per hour salvage pump on board, was engaged to proceed to Scarinish from the Clyde and, after patching the bottom and pumping out the flooded engine-room, tow the casualty back to Greenock for inspection and repair.

In view of the fact that no salvage pumps were immediately available at Tiree, it might have been wiser to have deliberately beached the damaged vessel on a flat, sandy bottom until *Warrior* arrived with its salvage pump and the engine-room could be de-watered and the bottom patched. For whatever reason, this course of action was not adopted and, instead, it was decided to keep the casualty 'safely' berthed alongside the pier until *Warrior* arrived. Unfortunately, almost as the tug came into sight the engine-room bulkhead, which had probably been damaged during the grounding, gave way and the vessel progressively flooded throughout. The MacBrayne's representative, and Hugh MacLeod, the salvage surveyor, could only watch in horror as the ship slowly settled onto the bottom before their very eyes. At high tide only the boat-deck, bridge and funnel remained above water. This unfortunate turn of events was almost a repeat of the *Raylight* incident two months earlier.

Considering the eventual outcome, the following report in Lloyd's Casualty List of 27 March 1973 appears to be a fairly casual assessment of events.

Oban 23 March

Motor vessel Loch Seaforth *which went aground yesterday near Tiree, sank today alongside the pier at Scarinish, Tiree. No one was aboard. The Chief Steward and Engineer had spent the night on board but left early today when they heard a rumble and the ship started to move. The vessel was due to leave for the Clyde towed by the tug* Warrior *which arrived just as she went down.*

Coastguard.

The pier at Scarinish is the only berth on the island of Tiree and with *Loch Seaforth* sunk alongside it the pier was completely out of action. This made it necessary to airlift in and out of the island airstrip, or ferry to and from the beach, using suitable small craft, all of the island's essential supplies and equipment. This state of affairs caused considerable inconvenience to the islanders and untold embarrassment to both MacBrayne and the Salvage Association.

The contract to remove the sunken ship was awarded to the Southampton-based salvage company, Risdon Beazley Marine. Using a heavy-lift floating crane they lifted *Loch Seaforth* and beached her well clear of the pier, where the bottom was then patched and the flooded spaces pumped dry before the vessel was towed to the Clyde for repair.

This whole episode caused Tiree's one and only pier to be out of action for a total of 53 days. It had been Bill Blackburn's worst nightmare so you can see why he was anxious that nothing would go wrong with the *St Rognvald* salvage operation up in Orkney. With his words of advice and warning ringing in my ears I left his office to catch the flight to Kirkwall.

The Orkney Islands lie north of the Scottish mainland, on Latitude 59 degrees, beyond the treacherous waters of the Pentland Firth, where the spring tides run faster than anywhere on the British coast. The Orkney archipelago is a group of more than 70 islands of which about 20 are inhabited. The largest of the islands is called Mainland; to Orcadians the Scottish mainland is simply 'Scotland'. The islands contain much evidence of prehistoric occupation: underground houses, standing stones, and earth houses. At Skara Brae in west Mainland there is one of the most complete relics of the late Neolithic Period that has ever been discovered. Norse raiders arrived in the late 8th century and colonised the islands in the 9th century; thereafter they were ruled by Norway and Denmark. The islands, together with Shetland, passed into Scottish rule in 1472 in compensation for the non-payment of the dowry of Margaret of Denmark, Queen of the Scottish King, James III.

The main town and port on Orkney is Kirkwall, located on a narrow strip of land dividing east and west Mainland. Daily ferry services link Kirkwall with Shetland and Aberdeen. Exports include lobsters, cattle and a very fine single malt whisky from the Highland Park distillery outside Kirkwall.

The origins of whisky distilling in Orkney are lost in the mists of antiquity but by the late 1700s there were numerous illicit stills hidden away from the excise men among the many little islands that make up Orkney. Around 1790 one of most successful of the 'Moonshiners' was a well-known preacher, Magnus Eunson, who spent his days at the kirk and his nights on the clandestine operation of several stills. I would like to think that the kirk evening service was preceeded by a 'happy

131

hour'. I'll bet there was no problem with kirk attendance in his parish! Some time in 1798 he came across a spring providing water of exceptional purity just south of Kirkwall, and promptly relocated his whisky production to take advantage of this superior ingredient. The Cattie Maggie Spring, as it is called, is located on what is now the Highland Park Estate where the fabled Highland Park Single Malt has been produced, using a distilling process that has hardly changed since the days of the Reverend Mr Eunson, for more than 200 years. The New York Times has described it as having '... *flavours that enchant the senses*'. Their gourmet food and drink critic does not exaggerate.

The approach to Kirkwall Bay, for all but small craft, is from the east by Shapinsay Sound, which is the stretch of water between the south side of the island of Shapinsay and Mainland. The String, a channel about half a mile wide at the western end of the sound, leads into Wide Firth past the north end of Thieves Holm Island. Kirkwall Bay lies at the south-east end of Wide Firth.

Admiralty Sailing Directions advise to '... *keep to mid channel through Shapinsay Sound and The String, clearing Thieves Holm Island by at least 2 cables on the approach to Kirkwall Bay*'.

They further advise that '... *the area contains many hazards and lacks prominent shore marks, and, since visibility is not infrequently poor, a degree of caution is necessary when navigating through the Sound and in Wide Firth*'.

St Rognvald was reported to have left Kirkwall for Aberdeen at 5.00p.m. on 4 May 1973. On board were 14 crew, 5 passengers, 204 cattle, 90 sheep, 2 goats and numerous barrels of Highland Park malt whisky. After clearing Shapinsay Sound and reaching open water, heavy seas and gale force winds caused a shift of cargo and the partial collapse of cattle pens in the 'tween decks. The vessel hove to while cargo was re-stowed and the cattle pens re-secured. The

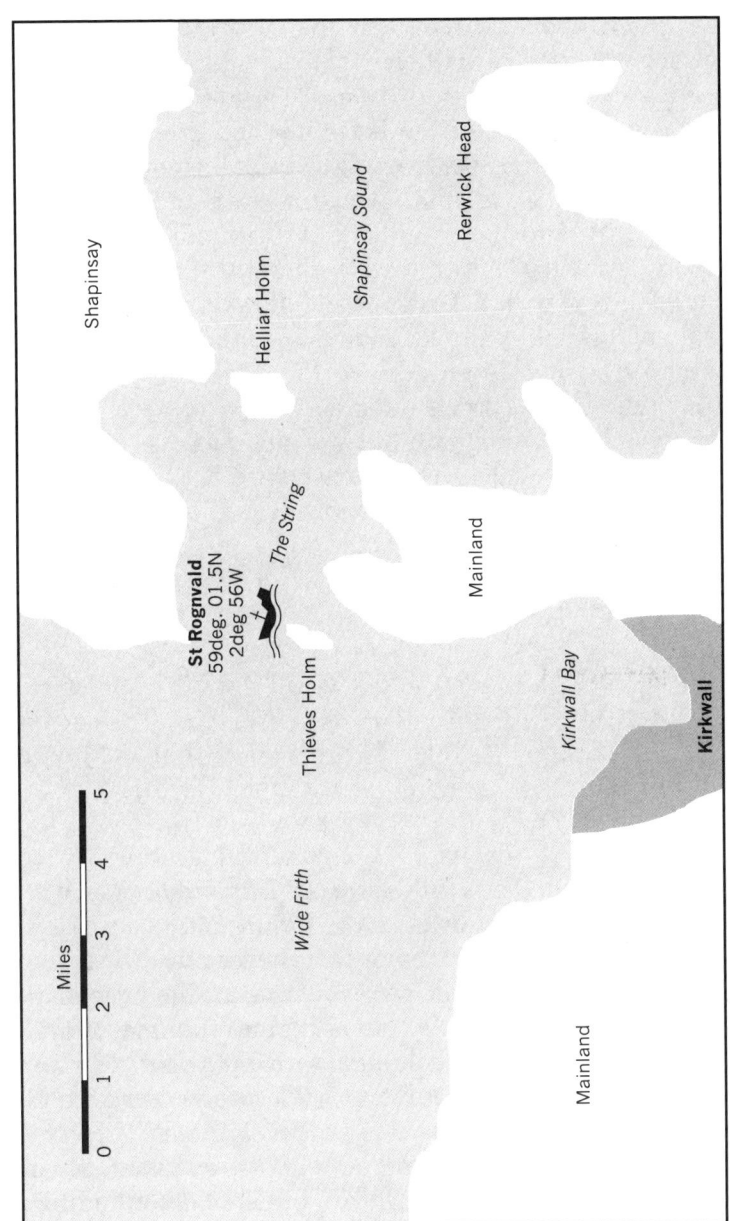

Map 9 Approach to Kirkwall

Miles

0 1 2 3 4 5

Shapinsay

Helliar Holm

Shapinsay Sound

Rerwick Head

St Rognvald
59deg. 01.5N
2deg. 56W

The String

Thieves Holm

Mainland

Wide Firth

Kirkwall Bay

Kirkwall

Mainland

master then decided to return to Kirkwall and await an improvement in the weather before resuming the passage to Aberdeen.

Based on a review of the log book and other records, it would appear that, while proceeding back through Shapinsay Sound and The String, the effects of wind and tide set the vessel well over to the south side of the sound. This would have brought them into line with the north end of Thieves Holm Island. Visibility was poor and, thinking that they were in mid-channel, they proceeded at full speed towards Wide Firth, only to run slap into the north end of the island.

Immediately after the grounding an attempt was made to refloat, using main-engines, but without success. The vessel was hard aground. Kirkwall Coastguard was informed of the situation and the local lifeboat was dispatched to the scene to take off the passengers. The vessel was aground on fairly-steep shelving rock, but while the forward end was almost completely dry, there was fairly deep water round the stern. This allowed small craft to come alongside abeam No. 2 hold later in the day and take off the cattle and sheep. In addition to off-loading the livestock, barrels of whisky stowed at the forward end of No. 1 hold were moved to the aft end of the cargo space in an attempt to lighten the fore end as much as possible.

Once the livestock had been removed ashore and the cargo trimmed aft, the Hull-based tug *Superman*, which had been engaged by the owners on £700 per day, was connected to the stern. An attempt was then made, during high water on 7 May, to tow the casualty clear of the ground, the tug being assisted by the main-engine running on 'full astern'. After some three hours, with no apparent movement astern towards deep water, and with the tide falling, the attempt was discontinued.

On 8 May the services of *Superman* were dispensed with and the tug sailed for Hull. On the recommendation of Bill

Blackburn, who, as mentioned earlier, had carried out a preliminary inspection of the stranded vessel, together with the owners' technical director, the owners invited professional salvage companies to inspect the vessel and submit competitive quotes for refloating.

Details of the salvage quotations subsequently received were as follows:

Bureau Wijsmuller Towage & Salvage Co.

Lightening vessel as required, patch leakages and refloat vessel. Replace cargo and equipment and deliver vessel safely afloat at Kirkwall. Lump sum Dutch Guilders 328,000. No Cure – No Pay.

Smit Tak Salvage & Transport UK Ltd

Services offered for refloating *St Rognvald* and delivering her safely afloat alongside quay at Kirkwall. Lump sum Dutch Guilders 250,000. No Cure – No Pay. Subject to contract.

After discussions between the owners and the Salvage Association, the salvage contract was awarded to Smit Tak Salvage & Transport UK Limited on 11 May. At this meeting I met the owners' technical director, Willie Ross. He had just returned from Kirkwall and, as he was a salvage legend – in his own mind at least – he gave me the benefit of his knowledge. He ended up by telling me, 'I know the people up there, Mr Walker, and I know the waters. Should you need any help, just speak to Mr MacKenzie, the manager of the Kirkwall Hotel – he is a personal friend of mine'. Exactly what advice on refloating the stranded vessel I could expect to receive from the hotel manager was not immediately obvious.

Smit Tak immediately began assembling their salvage

craft and the salvage vessel *Bever* arrived at Kirkwall on 14 May and anchored close to the casualty. The same day I was told to proceed to Kirkwall to look after the underwriters' interests and liaise as required with both Smit Tak and the owners.

I arrived at Kirkwall at about noon on 15 May to be met by Bert Scott, the owners' manager, at Aberdeen, and John Simpson, the technical superintendent. They whisked me off to the Kirkwall Hotel and brought me up to date with the situation.

They reported that Smit Tak had commenced salvage operations that morning with divers carrying out an inspection of the sea-bed around the casualty. Work had also started, laying out beach gear anchors on the seaward side of the grounded ship. In addition, a shallow draught vessel was now secured alongside No. 2 hold and work had started, discharging barrels of whisky.

On completion of this briefing the three of us went down to the quay where a boat was waiting to take us out to the casualty. Claus Reinecki, the Smit Tak salvage officer, was also going out with us. I introduced myself but was aware that he did not appear too friendly. 'I wonder what's wrong with him,' I thought. After arriving on site I had a good look round, together with Scott and Simpson. Reinecki wandered off on his own; he did not appear to be interested in showing me what his men were doing, which was strange.

St Rognvald was hard aground on steep, smooth-shelving rock on a heading of 230 degrees. She looked for all the world as if she was berthed on a slipway, with the hull, from the forefoot back to the forward end of No. 1 hold, completely out of the water. The depth of water then progressively increased to 5 feet at the aft end of No. 1 hold, 10 feet at the middle of No. 2 hold and 24 feet at the stern, these depths being recorded at high tide.

While the rock shelf on which the casualty was sitting was generally very smooth, there were several large, jagged rocks under the port and starboard sides of No. 1 hold. Contact with these rocks had caused damage to the bottom plating in that area.

On board the casualty all main and auxiliary machinery was fully operational and soundings of all the tanks and void spaces showed no signs of leakage. There was a very slight oil slick on the surface of the water round the stern. This was probably due to a leak at the propeller shaft seal.

Once I had completed my inspection I popped up into the mess-room for a coffee with the two owners' reps, Scott and Simpson. The Smit Tak salvage officer was in the mess but did not join us. He was busy reading his newspaper and looked as surly as ever. Over our cup of coffee I started to tell my two companions of my adventures on Sherbro Island six months before. When I mentioned Leen Breukel's name our silent, surly salvage officer in the corner suddenly came to life and put down his newspaper. He came over to where we were sitting and said to me.

'So you are the salvage guy who was in the African jungle with old Breukel. He is a good friend of mine – he told me all about it. Well, fancy that. Look, I must apologise for being rude to you when we came out on the boat, but I've had all sorts of experts up here these last few days telling me how to do my job. The worst was that owners' guy, Ross – one of those 'shake and bake' instant salvage experts who knows everything about salvage work and nobody else knows anything. I was afraid you would be just the same. But Breukel said you were OK. Now let's all have another coffee and talk about how we are going to re-float this ship.'

The agreed salvage plan was simply to remove as much

weight as possible from the vessel. This would mean discharging all the cargo, most of the fuel oil, fresh water and ballast. The protruding rocks under No. 1 hold would be cut away using pneumatic drills. At the same time, beach gear anchors would be laid out astern, ready to be connected to purchase blocks on the salvage vessel. Once all possible weight had been removed an attempt would be made to slide the casualty down the sloping rock shelf by heaving on the beach gear anchors and running the main engine 'full astern'.

Once the plan of action had been settled Scott, Simpson and I returned to Kirkwall in the work-boat. It was a bright, sunny day but there was a cold easterly wind and the three of us crouched in the little cabin to try and stay out of it.

At one stage the boat man said to me. 'I got a present from a fellah the day.' He produced a brown paper bag containing a half bottle of Bell's whisky which was then offered to me. In Western movies we have all seen bartenders sliding bottles of scotch along the bar-top for John Wayne, or someone of that ilk, to grab and down a slug of whisky straight from the bottle. But it is not so easy; I tried, but found that it is a bit difficult to deal with a mouth full of neat whisky. When I tried to swallow it some went down the wrong way and I felt as if most of it had gone up my nose. I was spluttering and choking all the way back to the Kirkwall Hotel. It's a trick I have never tried since. The Hollywood scotch bottles must contain cold tea, or else these film stars are 'real' tough guys.

Throughout 16 and 17 May the salvors were busy discharging cargo into their salvage vessel *Bever* and then transferring it ashore. To speed up the transfer of cargo from the casualty to Kirkwall quay, two locally-owned landing craft were chartered to assist *Bever*. Work was commenced, cutting away the rocks under No. 1 hold and preparing the blocks, sheaves and deck lashings needed for the two sets of beach gear anchors laid out astern of the casualty, ready to be connected to purchase blocks on the *Bever's* deck.

On the morning of 18 May all ballast was pumped out and the transfer of diesel oil from the No. 4 port and starboard double-bottom tanks and the cross-bunker tanks to *Bever* was commenced. Most of the fresh water was also transferred to the salvage vessel.

On returning to the hotel on completion of our day's work we were met in the foyer by MacKenzie, the manager. While we were chatting somebody said there was a call for us on the phone. Bert Scott was nearest and he took the call, which was from his boss, Willie Ross, our 'salvage expert', checking up on how we were getting on.

'Is that that b**t**d, Ross?' said MacKenzie, rather apprehensively. 'He's not coming back up to Kirkwall, is he?'

'What's that, Mr Ross?' said Bert on the phone. 'Oh, that was just Mr MacKenzie asking after you very kindly, Mr Ross.' And he calmly continued the conversation. It was obvious that Bert's ability to think fast in moments of crisis was second to none.

By 10.00p.m. on 18 May all preparations were complete and we were ready for a refloating attempt at high tide, which would be at midnight.

The details of the next few hours were recorded as follows in the official Salvage Log:-

22h45   Bever *moving away from casualty to a position some 3 cables off the starboard quarter and connecting to beach gear anchors.*

         Bever *connecting towing hawser to casualty.*

23h00   Bever *heaving up taut on beach gear and towing hawser.*

Various main engine movements ahead and astern together with full port and starboard helm movements.

Casualty heading swinging between 220 and 236 degrees.

No movement towards deep water apparent.

23h35   *Full astern on main engine.*

Bever *towing astern and heaving on beach gear anchors.*

23h38   *Casualty moving astern towards deep water.*

23h40   *Casualty afloat.*

23h45   *All tanks and bilges sounded. No signs of leakage detected.*

Commenced heaving in port and starboard anchors.

19 May 1973

Wind – south east force 3/4. Overcast with slight drizzle.

00h15   *Beach gear anchors disconnected.*

Commenced re-ballasting double-bottom tanks.

Bever *still connected aft.*

00h25   *Port and starboard anchors secured.*

00h30   Bever *disconnected and casualty proceeding towards anchorage under own power.*

01h15   Casualty safely anchored at Kirkwall Bay with
Bever *secured alongside.*

As soon as it was daylight the divers inspected the underwater hull for possible damage but found nothing serious – only scattered indents and scores under No. 1 hold and the engine-room. Later in the day, the previously offloaded diesel oil and fresh water was transferred back from *Bever.*

The salvage operation was now over, and on the morning of 20 May, *St Rognvald* sailed for South Shields, where the owners had arranged for her to dry dock for a detailed inspection of the underwater hull and to carry out whatever repairs proved to be necessary.

MacKenzie, the manager of the Kirkwall Hotel, had come out with us on the evening of 18 May, together with several of his staff and friends, to watch the refloating – or 'to watch the fun' as they put it. Once it was all over and *St Rognvald* was safely anchored in Kirkwall Bay we all went ashore together about 2.30a.m. On arriving back at the hotel, MacKenzie, in complete nonconformity with the Scottish Licensing Laws in respect of the sale and consumption of alcoholic beverages, opened the bar, and, to the cheers of the thirsty salvage team, declared, 'All drinks are on the house.'

Needless to say, mine was a large 12 year-old Highland Park, with a wee sensation of water just to emphasise the subtlety of the flavours.

Now, that is what I would call the right way to finish off a salvage job!

7

A Mystery of the Coast

Towards the end of 1974 I was transferred to the Salvage Association office at Durban in South Africa. This was a most interesting post where, in addition to the work in the Durban repair yards, we were involved in a wide variety of marine casualties up and down the African coast. Our 'beat' extended from Walvis Bay and the Skeleton Coast in what is now Namibia, but then known as South West Africa, to the Kenyan port of Mombasa, and included the Indian Ocean islands of Madagascar, Mauritius, Reunion and the Seychelles. During my time in the Durban office I attended numerous interesting and unusual casualties, but by far the strangest case concerned the Greek general cargo ship *Evdokia*. My involvement with this vessel came about thus.

Sunday 10 June 1979 saw me on my way from Durban to Port Elizabeth to attend on board the Algerian tanker *Wahran*. This vessel had sustained steering gear damage whilst on a ballast voyage from Trinadad to the Persian Gulf and had deviated into Algoa Bay to carry out repairs. Weather conditions at the time were foul, with southerly gale force winds and heavy seas lashing the coast from Natal to the Cape. Having arrived at Port Elizabeth in the early afternoon I travelled out to the tanker by launch, together with the British representative of the owners. Although Algerian owned, the vessel was managed by a UK based ship-management company and all the officers and technical management staff were British. A very heavy swell was running in Algoa Bay and we had a major problem transferring from the launch to the tanker. It took several attempts before the pair of us managed to grab the pilot ladder and scramble up onto the deck. This experience made us decide to remain on board overnight and wait until the weather improved before attempting to disembark. We had plenty to keep us occupied with the investigation of the steering gear problem. The manufacturer's engineers with the necessary replacement parts were not due to arrive from Greenock until the following day so it would be convenient for us to wait on board and discuss the situation with them when they arrived.

After dinner that evening I strolled up to the bridge to pass the time with the watch keeper and find out the latest weather forecast for the next day or two. On arriving in the wheelhouse the 3rd mate greeted me with, 'Looks as if there is another job for you coming this way.'

'What do you mean?' I asked.

'There has been a distress call from a Greek ship and she is fairly close; I expect she will make for Algoa Bay and should be here sometime tomorrow,' he explained.

The ship in question was the 7,144 gross ton Greek cargo vessel *Evdokia*, owned by Nicopan Shipping Co. of Piraeus and built by Harima Zosensho at Aio, Japan, in 1956. She had sailed from Durban on the afternoon of 8 June bound for Brazil with a cargo of copper ingots, paper rolls and asbestos. On 9-10 June the gale force weather that was sweeping the whole coast was encountered and, in the afternoon of the 10th, ingress of water was detected at No. 5 hold. The situation on the evening of 10 June is best described in the message received from the ship by Cape Town Radio.

143

MV Evdokia, Durban for Brazil, fully loaded, at 21.00 June 10 in position latitude 34 degrees 16 minutes south, longitude 24 degrees 18 minutes east course 270/275 degrees, needs assistance due water flooded in No. 5 hold, which pumped presently by vessel's means. Vessel is under strong gale proceeding Cape Seal.

Her reported position was about 90 miles west of Algoa Bay and some 50 miles from Plettenberg Bay, where they appeared to be heading, Cape Seal being at the western end of Plettenberg Bay. The situation did not appear to be too serious, but I wondered why they were heading for Plettenberg Bay; I would have thought that Algoa Bay was the obvious place to seek shelter and assistance. However, maybe the prevailing weather conditions made it easier for them to run for Plettenberg Bay. Anyway, on *Wahran* there was nothing we could do so I turned in for the night, leaving the situation to resolve itself in the morning.

The radio reports next morning were that *Evdokia* was still heading at slow speed towards Plettenberg Bay. In response to the master's request for urgent assistance a salvage tug had left Cape Town and South African Diving Services were preparing to airlift salvage pumps out to the vessel by helicopter. In addition, the tanker *Jastella* was standing by the Greek vessel, ready to assist if required.

After spending most of the day investigating the steering gear damage on *Wahran* and agreeing with the owners and the manufacturers on the best course of action, I went back up to the bridge in the early evening to catch up on the latest news on the Greek vessel. A Reuters message of 11 June describes the situation:-

Port Elizabeth 11 June

Airforce helicopters today lifted 16 crewmen from

Evdokia, *battered by heavy seas and listing badly off the South African coast. Port Elizabeth maritime officials said the vessel's Master and the remaining six crew stayed on board to sail the vessel to port. Maritime officials said a 50 knot gale was raging but if* Evdokia, *carrying general cargo,shipped no more water, she could reach the coast by late tonight.*

Reuter

Later in the evening we picked up further radio messages that advised that the salvage tug had reached the last reported position of the Greek cargo ship but had been unable to establish contact.

Then a message from the *Jastella* reported that *Evdokia* had increased speed, altered course away from Plettenberg Bay and was now heading east in the general direction of Algoa Bay. *Jastella* further advised that she was resuming normal passage. Based on the information received by radio and the last reported position of the distressed vessel, I retired for the night, confidently expecting, in spite of the continuing bad weather, to see the Greek vessel safely anchored close by in Algoa Bay in the morning.

First thing in the morning of 12 June I popped up onto the bridge to see what the latest situation was regarding *Evdokia*. There was no sign of her and no messages had been received from her during the night – all very odd. The situation had now taken a sinister turn. The South African Air Force had commenced an aerial search at first light along the coast between Plettenberg Bay and Algoa Bay. There was nothing to do but await the results of the search.

By now I had been contacted by my boss in Durban, Ian Lloyd, who told me that as soon as I was finished on *Wahran* I should get myself ashore; then, after contacting the representatives of the Greek owners of *Evdokia*, who

145

were due in Port Elizabeth that evening, find out what exactly the situation was and what could be done to protect the underwriters' interests.

The weather had now abated somewhat and there would be no problem disembarking by launch, so, as the steering gear repair was now in hand, I left *Wahran* in the afternoon and went ashore to the hotel to await the arrival of the *Evdokia* representatives. However, before leaving *Wahran* we picked up a message that one of the search aircraft had sighted wreckage near the coast, about 38 miles west of Cape St Francis. Later in the afternoon, a further message confirmed that the wreck of a vessel had been sighted on the rocks in the same position. Shortly afterwards, a further report was received that a survivor had been sighted on a rock and plucked to safety by an Air Force helicopter.

The gentleman representing the Greek owners turned out to be Gerald Geddes, a well-known London marine consultant. Next day, 13 June, the pair of us travelled down to the reported site of the wreckage, now confirmed as *Evdokia*, some ten miles east of Storms River Mouth. South African Diving Services were also going down to have a look, so we all travelled together. The drive from Port Elizabeth took two and a half hours. The first part was along the main coastal road for about 40 miles until we reached the eastern end of the Tsitsikamma National Park. We then left the main road and drove along a series of dirt roads in the National Park to reach the coast at Oubosstrand. The park is a forest and marine reserve running for 50 miles along the coast between Cape St Francis and Plettenberg Bay. Most of the coastline consists of spectacular rocks and steep cliffs, forming an awesome scenic seascape – a delightful vista for tourists, but it is a coast that ships normally keep well clear of.

Admiralty Sailing Directions in the Africa Pilot, Volume 3, advise that:-

Great care is necessary when navigating between Cape Seal and Cape St Francis due to the frequency of onshore sets. Keep at least 2.5 miles offshore by day and maintain depths of more than 45 fathoms at night or in thick weather – in places the cliff lined coast is fringed with rocks above and below water, extending up to 3 cables offshore. This stretch of coast is notorious for the number of vessels wrecked on it.

As we drove through the park, Gerald Geddes and I wondered what on earth *Evdokia* had been doing close inshore in such an inhospitable area and in such bad weather.

From Oubosstrand we travelled west through the Forest National Park, parallel with the coast-line, for some three miles and then reached Jaftaskraal, a cluster of houses and a little country store. We bought some cold drinks and chocolate bars at the store and, in answer to our enquiries about the shipwreck, they gave us directions to follow. Several police land rovers had gone along the track earlier in the morning and there had been considerable helicopter activity overhead, both that morning and the night before, so they had a good idea where the wreck was located – about a couple of miles to the west, they reckoned.

After fifteen minutes of bumping along the forest track we emerged close to the cliff-tops. Some distance away we could see police vehicles parked on a grassy slope near the cliff edge. This was the place. We parked close to the other vehicles, scrambled to the edge of the cliff and gazed at the waves crashing onto the rocks below.

We were adjacent to a gully that cut into the cliff-face for some 200 metres. The cliffs were more than 150 metres in height and almost perpendicular, with large, half-submerged rocks at sea level, on which the incoming swell pounded ceaselessly. Despite the tragic circumstances we couldn't help being aware that this was truly a most beautiful spot.

It was a bright and sunny South African winter's day with, to seaward, the sparkling, blue water of the southern Indian Ocean and the white, foaming surf breaking over the rocks and surging around the foot of the cliffs. On the shoreward side were the green trees of the forest and, some ten or twelve miles away, visible above the trees, the snow-covered 5,000 ft high tops of the Tsitsikammaberge.

Approximately 200 metres from the bottom of the cliffs, in amongst the half-submerged rocks, large sections of wreckage could be seen sticking up out of the water. It was not immediately apparent what part of the ship they actually were. One section looked for all the world like the tail fin of a large aircraft. Then we recognised it. It was the bow section, sheared off at the collision bulkhead and sitting upside-down on the rocks. We could see one of the anchors, the port one, still secured in the hawse-pipe. It was, in fact, the sight of the anchor that made us first realise that we were looking at the bow with the forefoot sticking up in the air. Some 50 metres from the bow, was the entire midship house. It had been cleanly sheared from the hull and was, like the bow, sitting upside-down on the sea-bed. Next to it was what appeared to be a section of the double-bottom structure. None of our party – not salvage divers, marine consultants nor surveyors – had ever seen anything like it. The ship had literally been torn apart.

On checking our Admiralty Chart 2084, we fixed the location of the wreck at ten miles east of Storms River mouth and 39.5 miles west of Cape St Francis, in geographical position 34 degrees 02.5 minutes south, 24 degrees 06.1 minutes east.

For more than half an hour we gazed in awed silence at the swell breaking around the broken sections of what, only 36 hours before, had been a living ship. Out beyond, and to one side of the upside-down midship house, I could see the swell breaking over what at first I thought to be a half-

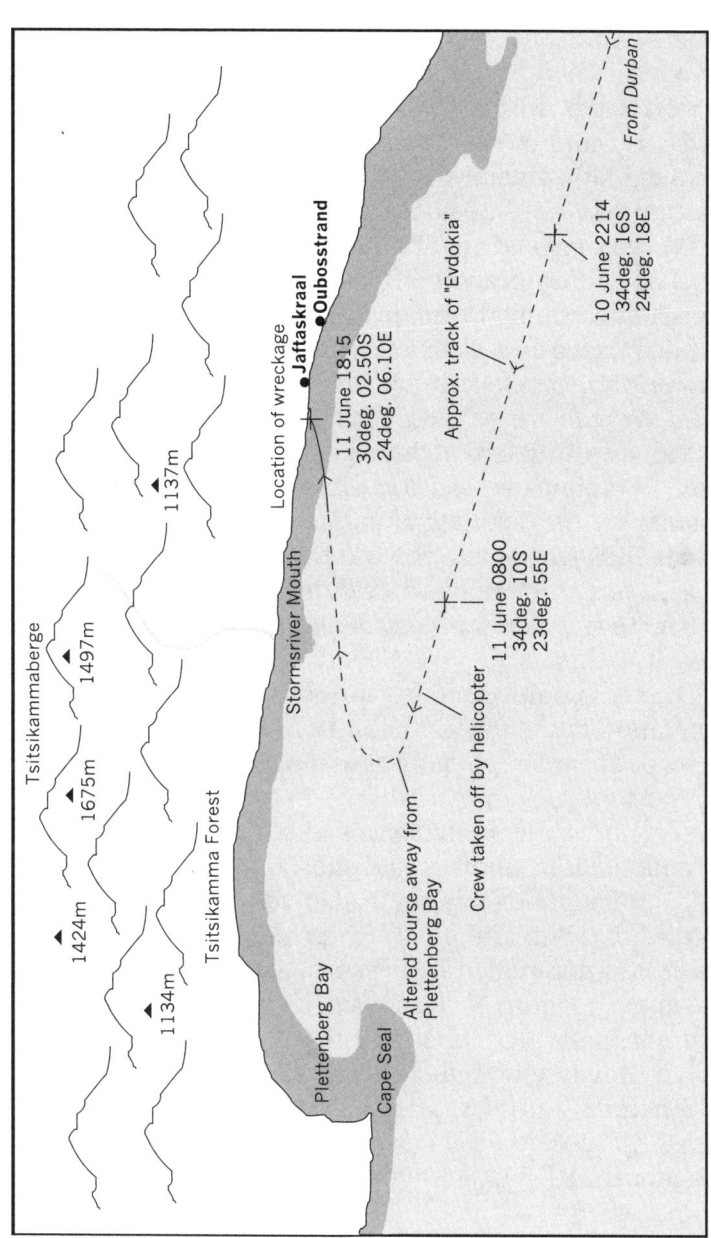

Map 10 Tsitsikamma coast, South Africa (From chart 2084)

149

submerged rocky ridge. However, closer attention showed that what I had taken to be rocks was, in fact, the six cylinder heads of the Sulzer diesel main-engine, clearly visible, now and then, in the troughs of the swell.

In the gully at the foot of the cliff a considerable amount of debris had collected, consisting of splintered dunnage, hatchboards, rolls of paper and asbestos. The South African Police were busy searching the gully and adjacent area for signs of the missing seamen. One body, that of the mate, was recovered as we watched.

There was nothing for us to do. We had seen all we needed to see – more, in fact, than we wanted to see. The drive back to Port Elizabeth was conducted in almost total silence. I think we were all wondering what the last moments on the doomed ship had been like. The sight of the scattered fragments of what had been the Greek cargo vessel *Evdokia* in amongst the rocks and breaking waves had been a terrible example of the awesome power of the forces of nature.

Next day Gerald Geddes was joined by a marine solicitor and the two of them were soon busy making arrangements to interview the crew members who had been lifted off by helicopter on 11 June; also the 3rd engineer who had been dramatically rescued from the rocks the following day. He was under observation in hospital and we would have to wait for a couple of days before anybody would be able to talk to him. While they were so engaged I returned to *Wahran* and completed my involvement in the steering gear damage.

The interviewing of the *Evdokia* crew started on Friday 15 June and lasted right through the weekend. The story that emerged was that, after leaving Durban on the afternoon of 8 June, everything had been normal, apart from the gale force weather conditions, until about mid-morning on 10 June when the chief engineer advised the master that there was leakage of water into No. 5 hold. The hold bilges had not been sounded. It would appear that on this ship the method of checking the hold bilge conditions was to start up the bilge pump and pump from each hold in turn until the pump drew air, indicating that the bilge was empty. The pump had been pumping from No. 5 hold for some considerable time with no signs of drawing air, so the conclusion was that the hold bilge was full of water. On being asked if a sounding of the No. 5 hold bilges was taken to confirm how much water was in the hold, the answer was: 'No, the weather was too bad'. Water was being continually shipped across the deck, making it dangerous for anyone to work out in the open. The general service and ballast pumps were all tried on the No. 5 hold bilges but there was no change to the pumping conditions. All pumps seemed to be pumping water from the hold; none showed signs of drawing air to indicate that the bilges were empty.

By the evening it would appear that the master and chief engineer were both convinced that there was a major ingress of water into No. 5 hold and the ship's pumps were unable to cope. At 10.14p.m. a distress call was transmitted, speed was reduced and the course altered so as to gain shelter in Plettenberg Bay in the lee of Cape Seal.

Throughout the night of 10-11 June the vessel steamed slowly towards Plettenberg Bay with various pumps working on the No. 5 hold bilges. In the morning, convinced that the flooding of the hold was serious, the master requested urgent assistance on the basis of Lloyd's Standard Form: No Cure – No Pay. Later in the day, the master requested that some of the crew be taken off. In response a South African Air Force helicopter evacuated sixteen of the crew about the middle of the day. This left six men onboard: the master, mate, chief engineer, 2nd engineer, 3rd engineer and the radio officer.

A press photographer was on board one of the helicopters during the crew evacuation and the photograph he took of the ship is interesting. It shows the vessel to be upright with no sign of the list that had been reported. It is not quite clear whether the ship is down by the stern or is pitching bow up into the prevailing sea-way. The aft deck in way No. 5 hold appeared quite normal – no signs of any heavy weather or other sort of damage to any of the deck fittings that could have caused water to enter the hold. Yet the master and chief engineer were apparently convinced that the No. 5 hold was flooding and the situation so serious that partial evacuation of the crew was necessary.

The only person who could tell us what happened after the sixteen members of the crew were airlifted from the vessel was the 3rd engineer. We were able to talk to him on the Sunday and he more or less confirmed what we had already been told about the events leading up to the helicopter evacuation on the afternoon of 11 June. One additional bit of information that he provided was that the strum box, or filter, on the bilge suction line from No. 5 hold had to be repeatedly cleared of debris during the 11th. This debris appeared to be mainly compressed and saturated paper. So there was water in the No. 5 hold, sufficient to reach the level of the paper rolls that were stowed above the copper ingots. At a rough calculation that meant there must have been at least 6 feet, or 2 metres, of water in the hold.

Following the departure of the helicopter he was down in the engine-room with the 2nd engineer. They were aware that the vessel altered course about 3.00p.m., at which time they were requested to increase the main-engine revolutions. Shortly afterwards, the chief engineer told them that the master was going to try and beach the vessel. As darkness fell, shortly after 5.00p.m. the chief engineer told them to set the main-engine revolutions at 'half-ahead' and then go up onto the bridge as the ship was closing on the shore and it might be too dangerous to remain in the engine-room. All six of the officers crowded into the wheelhouse as the ship steamed towards the coast. At approximately 6.15p.m. the ship struck the rocks close to the coastal cliffs and, almost immediately, started to break up.

'She hit the rocks and broke in two,' the 3rd engineer told us. 'The two halves just broke up and sank.'

As the bridge structure heeled over into the water five of the officers moved out to the starboard bridge wing where the crashing seas picked them off and flung them, one by one, into the foaming waves. One of them, possibly the master, was trapped in the wheelhouse and taken down with the ship. The 3rd engineer was dashed about amongst the waves and rocks and then, miraculously, one of the waves threw him up onto a rock and he was able to scramble a bit further up to a position of comparative safety. He remained, more or less uninjured apart from bruises and abrasions, huddled on the rock, cold and wet, all through the night. Of his companions there was no sign.

Shortly after daybreak an aircraft flew low overhead and he waved in a desperate attempt to attract attention. The aircraft was a Shackleton maritime reconnaissance aircraft, searching for wreckage and survivors, and they spotted him. In less than 20 minutes a helicopter was overhead; he was quickly winched up to safety and, in no time at all, was tucked up in a warm hospital bed – a very, very lucky man indeed.

On reaching the end of the crew interviews, and having listened especially carefully to the 3rd engineer's story, Gerald Geddes and I looked at each other for some time and said nothing. The whole thing just didn't make any sense. We were mystified.

The photograph of the ship taken from the helicopter on 11 June showed no evidence of its being in a near-sinking condition. There appeared to be no list although the photograph does suggest that she is trimmed by the stern, as would be the case with the No. 5 hold flooded, or partly flooded.

The operation of the pumps on the bilge system had apparently convinced the chief engineer that there was water in the No. 5 hold. The 3rd engineer's evidence regarding the presence of cargo debris in the bilge suction line would appear to support the chief engineer's assessment of the situation. However, there had been no attempt to confirm how serious the suspected flooding was by checking the hold bilge soundings. It also seemed very strange that distress calls were transmitted and an evacuation of most of the crew carried out without any attempt being made to enter and actually examine the conditions in No. 5 hold. The hold could have been entered via the access hatches in the mast house and an inspection carried out to ascertain if it was flooded or not. If there was water in the hold, then how serious was it? Were the pumps containing it or not?

The business of shipping water across the deck was no excuse for not thoroughly investigating the situation. Lifelines could have been rigged along the deck – after all, this is standard practice in bad weather when the crew are required to work or move about on an exposed deck. All in all, the available evidence suggests that no proper evaluation of the situation was carried out before requests for urgent assistance were transmitted and most of the crew evacuated.

The biggest puzzle of all, of course, was what was in the master's mind after the crew evacuation at noon on 11 June, when he altered course away from Plettenberg Bay and we all assumed he was finally making for Algoa Bay. Had the master maintained his original course for Plettenberg Bay he should have been able to anchor in the lee of Cape Seal by about 6.00 or 7.00p.m. that evening. Not as handy a location for assistance as Algoa Bay perhaps, but it would

would have provided a reasonable anchorage where the ship's problems, whatever they were, could have been dealt with – although, on account of the prevailing southerly gale, it would have been necessary to maintain the main-engine on immediate readiness in case the anchor started to drag. Instead, he apparently decided, for some unaccountable reason, to attempt to beach his ship on what is perhaps one of the most dangerous stretches of coast in southern Africa: a coast-line where, not even in one's wildest flights of fancy, is it possible to contemplate anyone attempting a beaching operation even in calm weather and in daylight – far less in gale force conditions in pitch darkness. There appeared to be no logical reason for this final, and fatal, decision. We could only assume that the master did not possess a copy of the Africa Pilot, Volume 3, containing the Admiralty Sailing Directions for this stretch of coast, which, as we have seen, clearly describe the dangers. Nor could he have studied the corresponding Chart 2084 that marks the location of all the offshore rocky and foul ground. Even the most casual perusal of these publications would have told the master that in approaching this section of the coast he was '. . . standing into extreme danger'.

We never found the answers to these questions. Only the master and chief engineer would have been able to provide them. But they were gone way beyond our reach.

8

A Case of Fraud

One quiet afternoon in the Durban office – it must have been about 20 January 1980 – our secretary, Karen, answered the telephone then told me that London office cargo department wanted a word with the boss, Ian Lloyd. He was out of town that day, so they had to make do with me instead. Their enquiry was whether we had had any dealings during the last few weeks with a ship named *Salem*, or had seen her recently in Durban. I looked through the 'case book', which listed all our surveys, for December '79 and January '80, but could see no mention of a vessel named *Salem*. Nor could I remember having seen a vessel of that name during my work around the port.

'What type of ship is she?' I asked, and was informed that *Salem* was a tanker, a 214,000 dead-weight capacity VLCC (very large crude carrier).

'Well,' I told them, 'if she is a large tanker she would have been at the Shell SBM (single buoy mooring) about one mile offshore, more or less abeam the airport, so she might well have been out there without us seeing her'.

'Her previous name was *South Sun* but we think it is possible that she may now have changed her name again to *Lima* or *Lema*,' the chap in the cargo department told me. 'Do any of these names ring a bell?'

'Oh yes,' I said, 'they certainly do. There was a tanker named *Lema* at the SBM just after Christmas.'

156

Are you sure?' they asked. 'This could be very important.'

'I am absolutely positive,' I told them. 'I saw her myself, a couple of days after Christmas, out at the SBM discharging cargo. I was out there at that time doing a minor survey on some superficial damage to the buoy. I could also make out the letters of her previous name – it was *South Sun*.'

After repeatedly seeking confirmation that there was no possible doubt in my mind about what I had seen they eventually rang off, sounding extremely pleased.

'I wonder what that's all about,' I thought after putting the phone down. But I had no doubts at all that I had seen a tanker named *Lema* out at the SBM on 27 December.

On that particular morning I had gone straight from home to the South African Diving Services (SADS) office on the Bluff at Durban. I then boarded their diving tender in order to proceed to the SBM where I was to conduct a survey of some recently sustained damage to various external fittings on the buoy. SADS were responsible for the operation and maintenance of the SBM and all surveys, including transport to and from the buoy, had to be arranged through them. The damage was not serious; the buoy was still operating normally and I reckoned that I would be back ashore by 12.00 or 1.00p.m. I had another two surveys to attend that afternoon, so was anxious to get back to the Bluff as soon as I could.

The tender set off down the harbour with me sitting in the wheelhouse, admiring the view, listening to the chatter on the VHF and chatting to the diving foreman, who was going to check, and agree, the extent of the damage and necessary repairs with me out at the buoy.

*

157

As soon as we cleared the breakwater I was surprised to notice that, instead of turning to starboard and heading south towards the SBM, we turned slightly to port and seemed to be heading towards the outer roads. The diving foreman, seeing my puzzled look, explained that we had to call at a Greek tanker before proceeding to the SBM.

'She will be berthing as soon as we have finished our inspection,' he said. 'But she has some problems with her deck machinery and we are going to put about a dozen line-handlers and riggers on board to assist them while connecting up to the SBM. It won't take long.'

By this time I could see the tanker; she was about three miles away. It was going to be well into the afternoon before I was back ashore. My plans for the afternoon's work were beginning to look a little shaky. The next thing that happened was that I overheard a message on the VHF to the diving tender. It was from the harbour pilot on board the tanker, advising that the master would be leaving the tanker at the anchorage and the chief officer would be taking over for the run down to the SBM.

'That's a bit funny,' I thought. 'I wonder why the master is in such a hurry to leave. I would have thought he would have stayed in charge until the completion of the berthing operation at the buoy. There must be some urgent reason for him to leave his ship at this stage.' My curiosity now aroused, I looked to see what the tanker's name was as we came alongside to transfer the line-handlers and riggers. On the starboard side of the bow, in what must have been the worst paint job I have ever seen, were what appeared to be the letters: L E M A. The same crudely painted letters were on the stern. Both sets of letters appeared to have been fairly recently painted. Underneath the welded letters of her original name could be seen: *South Sun*. 'Well,' I thought, 'whatever sort of a crew they've got on board, painting ships' name letters is not one of their talents.'

As soon as we had completed the transfer of the line-handlers and riggers, we set off for the SBM. The job there was fairly straightforward and, once we arrived, it didn't take long for the foreman and I to agree on what repairs were necessary and, by about noon, I had all the information I needed to write a survey report for the underwriters concerned. As I looked over my notes to make sure I had not missed out anything, I became aware that the pilot on the tanker – it was now only about one mile from the buoy – was calling us on the VHF, asking if we could take the master ashore. Apparently, he had not been able to get off at the anchorage and was now anxious to reach Durban before the banks closed!

'That's all we need now,' I thought. 'By the time we have waited for this guy and got him back to Durban the afternoon will be gone. My other surveys are going to have to wait until tomorrow. Anyway, where the hell is the ship's agent? He should be looking after the master, not us.'

We moved clear and waited for *Lema* to come up and be secured to the SBM. This, owing to the problems with her deck machinery, took about an hour or so. Then we moved alongside the tanker and waited for the master. After a further delay a youngish-looking man, later identified as one Demetrios Georgoulis, came down the accommodation ladder and boarded the diving tender. He had no luggage – only a very smart looking briefcase. I wondered if he was going to make a large deposit at the bank. On the other hand, maybe the crew had to be paid and he was going to make a substantial withdrawal. Whatever it was, it would have to wait until the next day; the doors of the bank would be well and truly closed long before we got back to the Bluff.

In actual fact, by the time we did berth next to the SADS office on the Bluff it was well past 5.00p.m. The *Lema's* master got himself a taxi and vanished in the direction of the city. Meanwhile, I got busy on the phone, explaining to those concerned why I had not turned up that afternoon to carry out their surveys as had been previously arranged. They had, of course, been standing about for most of the afternoon waiting for me and were not terribly amused. But as our National Bard once famously put it: 'The best laid schemes o' mice an' men/Gang aft agley.' They do indeed.

So you can understand why there were absolutely no doubts at all in my mind about seeing a tanker named *Lema* at the Durban SBM on 27 December 1979.

London's interest in this tanker had started on 17 January 1980 when *Salem* sank off Dakar in what was thought to be a fully loaded condition. The underwriters faced a possible $24 million claim on the hull and a further $56 million for the cargo.

The following casualty reports appeared in Lloyd's list of 24 January:-

London 18 Jan

The following Lloyd's radio telegram has been received from tanker British Trident *via Portishead Radio, timed 1632 GMT 18 Jan: Following is brief account of rescue of crew of tanker Salem; all times are GMT*
17 Jan: At 1050 vessel sighted starboard bow with unusual list and trim. At 1100 distress received by radio. At 1115 orange smoke sighted. At 1129 two lifeboats

160

sighted. At 1136 Salem *sank six cables on port beam*
British Trident *examining life rafts and debris. At 1224
first boat alongside. At 1231 second boat alongside. At
1235 all 24 survivors on board. 1235 to 1326 crew of*
British Trident *recover boats from water. Vessel proceeded Dakar at full speed, arrived at 2300 and cleared
at 2400.*

Dakar 19 Jan

Following obtained from Master of Salem*:* Salem *ex*
South Sun, *flag Liberian, owners Oxford Shipping Co.
Inc. Cause of explosion unknown by both master and
chief engineer. Time of explosion 0500, 16 Jan. Time of
sinking 1136, 17 Jan. Vessel loaded 193,000 tons crude
oil Persian Gulf for Italy but was proceeding Tenerife
for bunkers.*

Lloyd's Agents

Dakar 21 Jan

Salem *ex* South Sun, *registered Monrovia: Cause of
explosion unknown but occurred 0500, 16 Jan. Vessel
sank 1136, 17 Jan, in position latitude 12 degrees 38
minutes north, longitude 18 degrees 34 minutes west.
Cargo 193,000 tons crude oil. Master and officers are
kept in Dakar for official inquiry.*

Lloyd's Agents per Salvage Association

However, in the days that followed, the underwriters and
their legal advisors became more than a little suspicious of
the circumstances surrounding the sinking. Firstly, the BP
tanker *British Trident*, which saw *Salem* sink and later
picked up the crew from the lifeboats, observed that the
survivors had all their personal effects with them neatly

packed in their suitcases. Secondly, the crew's story that the sinking had been caused by a series of huge explosions was more in keeping with an empty tanker whose tanks, following discharge of cargo, invariably contain an explosive mixture of petroleum gas and air, rather than a fully loaded one. Furthermore, *British Trident* saw no signs of the fire that would surely have followed a major tanker explosion, nor the sort of massive oil slick that would have been expected to spread across the ocean in the area of the sinking, had *Salem* been carrying 193,000 metric tons of crude oil.

The *Salem* drama had apparently started several months before when a certain American-Lebanese businessman, later identified as one, Fred Soudan, obtained a contract from the South African Fuel Fund Association to deliver a cargo of crude oil to Durban. Barred by OPEC and UN sanctions from obtaining oil through normal channels, the South Africans were prepared to pay $43 million for the cargo. Armed with this contract, Soudan and his associates obtained an advance from a South African bank and with it purchased the 214,000 dead-weight tanker *South Sun*. This vessel, owned by Pimmerton Shipping Ltd. of Monrovia, was lying at anchor off Dubai, having arrived there on completion of a ballast passage from Quintero Bay, Chile, on 16 November 1979. On, or about, 30 November *South Sun* was sold to Oxford Shipping Co. Inc. – directors included Mr Fred Soudan and Demetrios Georgoulis – and renamed *Salem*.

The next move was to obtain a cargo. They were put in touch with an Italian company who had just bought 193,000 metric tons of crude oil from the Kuwait Government and, as a result, the Italians chartered *Salem* for a voyage to Europe. *Salem* proceeded from Dubai to Mina al Ahmadi and, after loading the cargo of crude oil, departed for Italy on 10 December. Following *Salem's* departure from Kuwait, the Italians sold their cargo to Shell, but delivery was still to be in Europe. Soudan and his pals were now set to sell Shell's cargo to the South Africans. Their plan was to change the tanker's name from *Salem* to *Lema*, discharge the cargo at Durban, then sail off towards Europe, change the name back to *Salem* and scuttle the ship at some suitable location off the North African coast. Shell would then be advised that their cargo had been lost as a result of a series of severe explosions that had sent *Salem* and cargo to the bottom of the ocean. Soudan and company would then be left to share the $43 million received from the South Africans.

In view of the suspicious reports from *British Trident*, plus that fact that nobody could find any trace of the sort of large oil slick that would normally be associated with the sinking of a loaded tanker, it looked to the underwriters as if the 193,000 metric tons of Kuwait crude had possibly been off-loaded prior to the sinking – but where? The international sanctions imposed on South Africa made the clandestine delivery of a cargo to that country more than a possibility and, for a tanker of *Salem*'s size, the only facility where she could discharge would be the SBM at Durban.

One of the problems about trying to trace tanker movements in South Africa was that, because of the OPEC and UN sanctions, oil imports had to be made in secret. Ship owners did not want it to be made known that they were delivering oil to South Africa in case they became the targets of legal action by their governments. For this reason the reporting of tanker movements in South African ports was banned under the South African Official Secrets Act. Therefore, the only way to find out if *Salem* had called at Durban would be if someone had actually seen her – someone not bound by the Official Secrets Act! This, then, was the reason for the call I received from London on 20 January 1980.

As the case unfolded the following reports appeared under marine casualties in Lloyd's List of 4 Feb and 11 Feb:-

London 1 Feb

It has now been confirmed that crude oil from Salem *was secretly discharged at Durban. Scotland Yard's Fraud Squad is now involved in investigating the circumstances surrounding the disappearance of the cargo, insured for $56 million, and of the tanker in deep water off the west coast of Africa. All cargo insurance was placed in London and a statement from Lloyd's late yesterday said cargo underwriters had been making an investigation through their solicitors. 'This had disclosed that the vessel called and discharged most of her cargo at Durban at the end of December,' the statement said. A Tunisian member of the crew, interviewed by a lawyer and diplomat in Paris, claimed that the tanker was deliberately sunk. He said that the bulk of the cargo had been unloaded in South Africa and the tanks filled with sea water to make* Salem *appear laden.*

Athens 5 Feb

The master of Salem *has been charged in connection with the sinking of another vessel a year ago, a spokesman for the Greek Ministry of Merchant Marine said today. He said Demetrios Georgoulis, master of* Salem *had been implicated in an investigation into the sinking of a vessel and the illegal sale of her cargo but a date had yet to be fixed for his trial. The spokesman said the vessel sank after being diverted to Lebanon where her cargo was sold, but he left Greece before Judicial Authorities could implement a ruling keeping him in the country.*

Once the facts of the case became known the South Africans paid Shell $30 million for the cargo that had been stolen from them. Owing to a problem with the ship's cargo pumps only 180,000 metric tons of cargo had been discharged at Durban on 27-28 December 1979, instead of the 193,000 metric tons loaded at Kuwait. The balance of 13,000 metric tons presumably went to the bottom with the ship on 17 January 1980. This amount was subsequently valued at $3.8 million and was recovered by Shell from the cargo underwriters. Hull underwriters, on the other hand, have never received a claim for the loss of *Salem*. In the United States, Fred Soudan was eventually prosecuted by the US Justice Department and received a substantial jail sentence. In Greece, a shipping agent and several of the *Salem* crew members, including, presumably, my friend, Demetrious Georgoulis, also received jail sentences for their part in the fraud.

One of the morals of this affair must surely be that, if you are engaged in a criminal operation like stealing $56 million worth of someone else's crude oil, you must play it cool. Do not, under any circumstances, arouse the curiosity of the odd bystander, who might well be a trained observer – such as a marine surveyor, for instance. Do not make unusual calls on the radio about the master handing over to the chief officer more or less in the middle of entering port. Do not make a 'pig's ear' of painting-in the change of the ship's name. Do not request our, by now very curious, 'bystander', to give the master a lift ashore so that he can get to the bank before it closes. All these things will conspire to ensure that the 'bystander' will remember the ship's name, as well as exactly where and when he saw her.

9

Europa Island

We recently moved house, having found that our original retirement home was far too big for us. It was an old house, full of character, but expensive to run, especially in winter, with its inefficient central heating system and lack of effective insulation. The move was only round the corner and up the road a few hundred yards, but it was every bit as chaotic as many of our previous intercontinental re-locations. The only part of the move that I enjoyed was sorting out my book collection, some of which I had not seen since leaving Singapore. In one box I came across a photograph album and an old hard-backed journal that I had not seen for years. The album contained photographs taken during a salvage operation on a stranded fishing vessel near Madagascar. The journal was the daily salvage log, which is reproduced at the end of this chapter. On my first day on the stranded vessel I had come across the book in a corner of the wheelhouse. It was a like an old ledger or cash-book of the type that used to be common in offices before the advent of computerisation. Only the first few pages had been used; they were filled with Korean characters and appeared to be lists – details of fish caught perhaps. I decided the book would make an ideal salvage log.

Twenty years have slipped past since I took the photographs and wrote up my log at the end of each day, but as I looked at my handwritten notes on the slightly yellowing pages, I was suddenly transported from a cold, wet Scottish winter to a sunny tropical island. Not the type of island paradise that is portrayed in the television holiday programmes – all golden sand, waving palms and lush vegetation. No – this particular island was very hot, very dry, very barren and with more rock than sand along the shore.

It was on Tuesday 8 April 1980 that we received a telex in the Salvage Association Durban office from Korean underwriters, advising that *Kwang Myung 156*, a Korean long-line tuna-fishing vessel, had grounded and been abandoned at Europa Island. The message instructed us to send a surveyor to the casualty location and ascertain the condition of the stranded vessel and the prospects of successful refloating. In the office our immediate reaction was: 'Where the hell is Europa Island?'

Charts were pored over and there it was – a tiny spot in the Mozambique Channel, 160 miles west of Madagascar on latitude 22 degrees 20 minutes south, longitude 40 degrees 22 minutes east. The Admiralty Sailing Directions in the South Indian Ocean Pilot give the following description:-

Ile Europa – Danger.

The island, which is very difficult to distinguish at night, is composed chiefly of sand, with low hummocks in places; it is partly covered with bushes and there are some trees attaining an elevation of 80 feet.

A flagstaff, two small radio masts and a white hut of the Meteorological Station are situated half a mile eastward of Pointe Nord-Ouest, the north-western tip of the island.

There is no secure anchorage off Ile Europa; it is possible however to anchor near the edge of the reef fringing the northern coast of the island. There is no room to swing in the event of a shift of wind and the holding ground is poor.

167

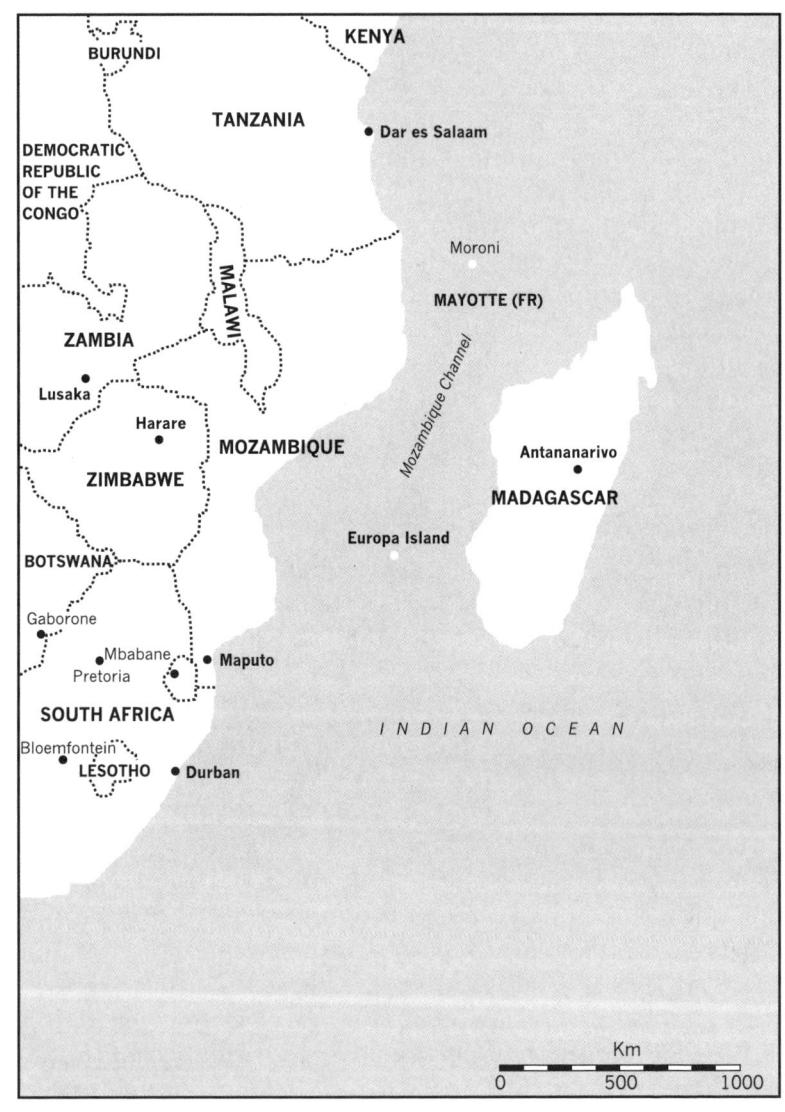

Map 11 Mozambique Channel showing Europa Island, Madagascar and South Africa

Further research revealed that the island was a French possession administered from Reunion, 1,000 miles to the east. The island has a land area of 28 sq. km. and a total coastline of 22.2 km. There is an airstrip located about one and a half kilometres south east of the Met. Station and the staff are flown in from Reunion and changed over every two weeks, together with a platoon-strength detachment of French troops who occupy a small camp adjacent to the airstrip. All supplies for the Met. Station personnel and the troops are airlifted from Reunion. There is little or no fresh water on the island. All things considered, it did not appear to be a very hospitable sort of place.

South African Diving Services (SADS), a Durban based salvage company, expressed an interest in the stranded vessel and started making arrangements to send an inspection team to the island with a view to ascertaining whether a salvage operation would be a viable commercial proposition for them. This interest from SADS solved the problem of how to get to Europa. There are no commercial flights to the island, only the fortnightly French military flights from Reunion and the prospects of being able to use one of those were almost zero. A charter flight with a Cessna 310 aircraft was arranged by SADS for the 700 mile flight to the island and I was told by my boss, Ian Lloyd, to join them.

The SADS's inspection team consisted of the boss, Monty McKeown, salvage engineer Graham Louw and their senior diver, Mike Smith. They brought along with them a Zodiac inflatable boat complete with outboard motor, sundry other items of salvage equipment and enough essential supplies, including fresh water, to keep the four of us for a week.

We left Louis Botha airport, Durban, at 8.00a.m. on 11 April and, after a refuelling stop at Maputo, landed on the airstrip at Europa at 12.30p.m., to a somewhat frosty reception from the French Army. Apparently, we should have

169

obtained clearance from the authorities in Reunion before leaving Durban. We had, in fact, tried to contact them but could not get any response, so we just went. A combination of Monty's charm and two bottles of Johnny Walker Black Label (SADS's medical comforts – they never leave home without an adequate supply) soon sorted out the business of our not having waited for official clearance to fly in. In no time at all half a dozen French paratroopers were helping us move our equipment from the aircraft to the Met. Station which, the paras advised, would be the best place for our base camp. There was more than enough room; in fact, it had been used as the base camp for a French geographical expedition several years before.

After dumping our personal kit in the space assigned to us, we spent the period until sunset getting all the equipment organised down on the beach so that we would be able to start for the stranded vessel first thing in the morning. The French troops reported that *Kwang Myung 156* was aground on the eastern side of the island and they recommended that we approach the casualty from seaward, proceeding north about round the island from the Met. Station.

We were up, bright and early, next morning. Monty produced breakfast and, after a few last minute adjustments to the Zodiac, we set off. The distance by boat round the north of the island was about eight miles. We passed inside the reef off the north coast, crossed the mouth of the lagoon, rounded the north-eastern point and then sailed down the east coast, again inside the reef, arriving at the casualty at about noon.

Kwang Myung 156 lay hard aground, bow on and at right angles to the shore, in geographical position 22 degrees 21 minutes south, 40 degrees 23.5 minutes east. It was almost high tide when we arrived and we were able to make fast to the stern and scramble on board. Initial inspection showed

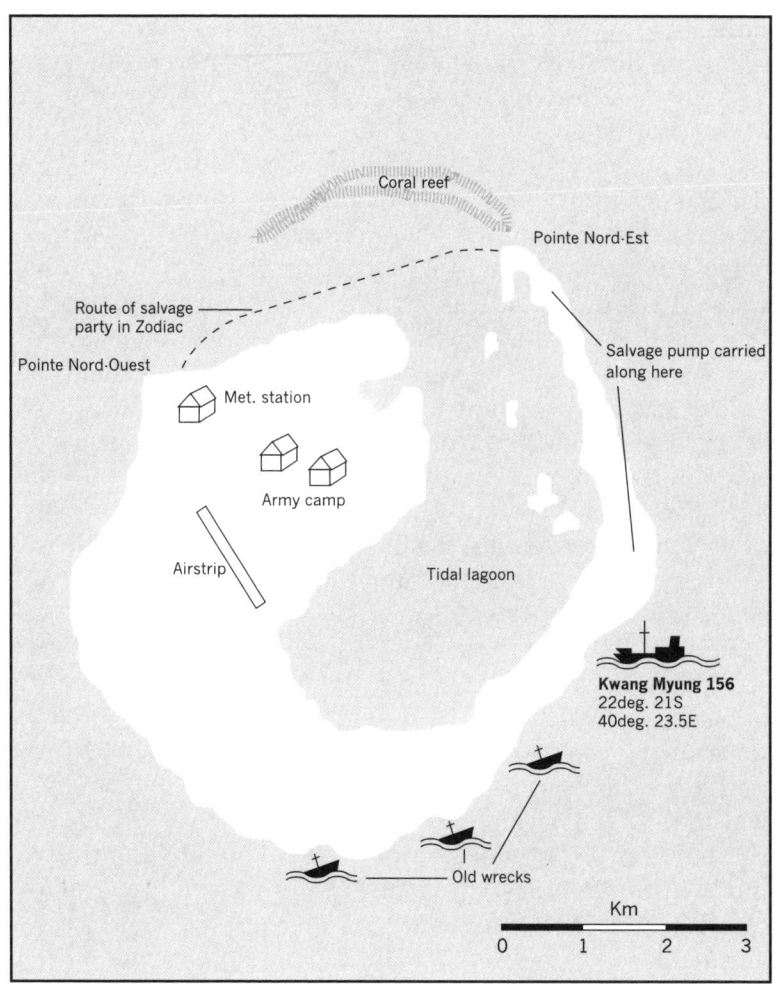

Map 12 Europa Island

171

that the No. 5 double-bottom tanks and the port aft coffer-dam were open to the sea but all other spaces appeared to be sound, with no evidence of ingress of water. The fish hold was full of frozen tuna still in good condition and the main and auxiliary machinery appeared to be undamaged.

At high water some movement could be detected at the bow but the aft half of the vessel appeared to be firmly on the bottom. The situation at low water was that the hull dried out almost completely and it was very nearly possible to scramble down a pilot ladder and walk ashore without getting your feet wet. Inspection of the underwater hull showed scattered indents, mainly in way of the aft half of the vessel. The rudder was hard over to port and the skeg was fractured at the forward end of the propeller aperture.

We returned to the Met. Station in the late afternoon, taking the opportunity on the way to catch some fish for our supper. The sea around Europa is simply teeming with fish. The water is crystal clear and looking down through a diving mask revealed a veritable kaleidoscope of brightly-coloured coral, exotic-looking tropical fish and turtles. It was a scuba diver's dream-world.

Back at the Met. Station our intention was to barbecue 'the catch of the day' down on the beach while we enjoyed a couple of cold beers. However, halfway through cooking the fish we came under attack from what appeared to be millions of the most ferocious mosquitoes imaginable. Such was the vicious nature of the onslaught, that we were forced to abandon the barbecue and retreat behind the safety of the Met. Station mosquito nets. The French soldiers and the weathermen had mentioned that mosquitoes were a prob-lem in the evening, but we did not pay much attention at the time. I have never in all my life come across such a nasty insect as the Europa Island mosquito. In comparison, the notorious West Highland midge, normally not to be taken lightly, pales into complete insignificance. Later, we

172

were astounded to learn that these mosquitoes were a protected species. They are the only mosquitoes in the world that breed in salt water. I am afraid we considered this to be a conservation step too far.

Next day, we returned to the casualty and carried out further inspections at all states of the tide. Soundings taken round the casualty, and on the seaward side, showed deep water some 50 metres astern. A further 100 metres to seaward, the bottom dropped away steeply to 200 fathoms (about 360 metres). Taking into consideration the condition of the stranded vessel, and the distance to deep water, we reckoned that refloating should not be a problem. The main difficulty would be the remote location; all equipment would have to be mobilised from Durban some 700 miles away.

Having by now obtained all the information we required, it was back to the Met. Station where messages were transmitted, via Reunion, to the owners and underwriters in Korea, advising details of the inspections carried out and that we considered successful refloating was possible. Then there was nothing more to do but return to Durban and wait while SADS put together their salvage plan and made a formal offer to the owners for refloating the stranded vessel.

On 21 April the owners and the underwriters accepted the SADS offer and work was immediately commenced, preparing salvage equipment for airlifting to Europa Island. At the same time, the anchor-handling tug *Causeway Salvor*, then at Cape Town, was ordered to Durban to embark the necessary heavy equipment before sailing for Europa Island.

The advance salvage party flew in to Europa Island on 22 April while I joined the follow-up party who flew in two days later. On 23 April Monty and the rest of the advance party travelled by Zodiac round the north of the island to the casualty location. On arrival they were disappointed to discover that conditions were not so favourable as they had

173

been during the preliminary inspections of the previous week. The salvage crew had to swim from the Zodiac through the surf to reach the stern and scramble on board. A check of the position showed that the casualty had moved about 50 metres up the beach from where the initial inspection had been carried out. Below deck they found the engine-room flooded to a depth of two metres at high water, submerging all the lower-level motors and the starboard auxiliary engine and alternator. The flooding was most probably the result of tearing of the bottom plating under the engine-room when the vessel was driven further ashore during a storm, which must have occurred after we departed for Durban on 15 April. Even at high water and with a moderate swell running there was no movement in the vessel.

The salvage operation was not going to be nearly as straightforward as originally envisaged.

Monty and his party repaired back to the Met. Station to await the arrival of the rest of the team and revise their salvage plan in the light of what they had found. It would not now be possible for the salvage team to transfer their salvage pump and associated equipment onto the casualty from seaward using the Zodiac. All the equipment would have to be carried overland and rigged on board at low water. Based on the changed situation, and taking into consideration advice from the French Garrison, it was decided that probably the best way would be to ferry the equipment, using the Zodiac, from the beach at the Met. Station to the north-east tip of the island (Pointe Nord-Est) where unloading could be carried out in sheltered water at the mouth of the lagoon. From there everything would have to be carried down the eastern shore of the island to the site of stranding.

I landed back on the island, together with the balance of the SADS team, on the afternoon of 24 April. During the day Monty and his advance party had carried out a further reconnaissance at the casualty location. In the evening the deteriorated situation was discussed but there did not appear to be any alternative to carrying all our equipment overland from the mouth of the lagoon. The general salvage plan was firstly to recommission the port auxiliary engine and alternator, which was apparently more or less unaffected by the flooding; then, with power available, discharge sufficient fish cargo, fuel, stores and fishing equipment to bring the vessel to her floating draughts. The possibility of laying out beach gear anchors in order to drag the vessel towards deeper water was discussed but considered to be impractical owing to the steep-shelving nature of the seabed to seaward of the casualty.

In addition to the original inspection team, the salvage team now contained Frank Mcloughlin, an ex-Safmarine (South African Marine Corporation) engineer, who was to be responsible for re-activating the casualty machinery. Now that the engine-room was flooded this was not going to be so easy. I had been involved with Frank on a previous case several months before and had not been too impressed with him on that occasion. I wondered if Monty had picked the right man; I had my doubts. Also included in the team were two additional divers and one mechanic/electrician.

Next morning, 25 April, all hands and the cook (Monty) turned to, shortly after daybreak, and commenced moving the salvage gear by Zodiac from the base camp area to the lagoon entrance at the north-east of the island. That was the easy part – then came the difficult part: carrying it down the shore to where *Kwang Myung 156* lay aground. The nastiest piece of equipment to carry over the rocks, coral and pools of the shore was the Honda petrol-driven salvage pump. After a considerable amount of experimenting, and much bad language, the pump was slung between two bamboo poles, which were carried on the shoulders of four good men, rather in the manner of a coffin.

175

The distance from the mouth of the lagoon to the casualty was only about three miles but under the weight of that pump it felt like thirty. The carriers had to be changed over quite frequently as they staggered through and over pools, rocks and coral. During my spells of helping to carry the pump I found that the bamboo pole dug painfully into my shoulder as we bobbed up and down over the rough ground. It was a most unpleasant business.

Such was our slow rate of progress that we failed to reach our destination that day. Frank and one of the divers continued on to the casualty with some of the more portable items of equipment. The rest of us secured the pump in a safe spot and returned to the Met. Station, just as the swift tropical sunset brought the working day to a close. At our camp we found Monty, who had remained behind that day, busy trying to contact Durban on his C.B. radio.

'I've been working this radio nearly all day and the only fellow I can raise in South Africa is a some guy in Oudt-shoorn in the Karoo of all places,' he told us – then added, 'He says he's the bloody local undertaker.'

'Just the very man we need,' was my response. 'Tell him to get six of his strongest pallbearers over here right away and they can carry this confounded Honda pump for us.' In my mind's eye I had an hysterical vision of six pallbearers, wearing top hats and dressed in black, solemnly carrying the Honda pump, draped in black crêpe, along the island shore.

We never did get the pallbearers but, until the tug arrived, our Oudtshoorn undertaker provided a most valu-able radio link with Durban. Every evening Monty would contact the funeral parlour and our friend would then relay messages to and from the SADS office for us. I wonder if he ever sent in a bill for his telephone charges.

On the morning of 26 April we returned to do battle with the pump. It proved to be just as heavy as it had been the day before; in fact, the weight of the pump seemed to increase in direct proportion to the distance carried. Nevertheless, step by step, hour by hour, we drew closer to our destination, finally arriving on the shore opposite *Kwang Myung 156* at around 5.00p.m., where we all collapsed, exhausted. On arrival we were greeted by Frank who, instead of telling us what a grand job we had done, informed us that he had lost his toothpaste – '. . . stolen by the hermit-crabs', he alleged. How the hermit-crabs got hold of his toothpaste we never found out.

Once we had got our breath back work was started getting the salvage gear on board, until the mosquitoes forced us to call a halt for the day and seek shelter in the wheelhouse. Fortunately, Monty managed to produce a meal before the mosquitoes struck. Exhausted by our exertions of the day, we slept the sleep of the just.

An inspection of the engine-room next morning at high water showed that, as a result of a starboard list of some five degrees, the starboard generator and all the pumps and motors on that side were completely submerged. The machinery on the port side, which of course was the high side, was generally clear of the water, except that the lower section of the alternator stator and rotor coils was in the water. As the tide dropped the water drained out of the engine-room and, much to my surprise, the SADS team prepared to start up the port generator.

'You can't do that!' I exclaimed in some alarm. 'The stator and rotor coils are wet – they've been under water.'

'Oh, that's all right,' said Frank. 'We ran it at low water last night so that we could have some lights and see what we were doing. Don't worry – it will be OK.'

I watched in amazement as the generator was started, run up to speed and the circuit-breaker closed. The alternator appeared to be quite unaffected by its contact with

177

salt water, but the diesel prime mover cooling pump lost suction after a while and the unit had to be shut down to prevent overheating.

With the vessel sitting firmly on the bottom it was not possible for the divers to get at the splits and holes that were causing the engine-room to flood on the rising tide. The plan was, therefore, to try and rectify the port generator cooling system so that it could be used to provide power for the bilge and general-service pumps, which would then be used to de-water the engine-room bilges and prevent flooding on the rising tide. At the same time, the deck winch would be used to remove whatever weight was necessary to bring the vessel to a floating draught. The most easily removed weight in the vessel was the fish-cargo, estimated at 75 tons. While it would be possible to discharge the fish by hand, without the assistance of the power driven winch, this would be a rather lengthy business. The availability of the port generator was therefore vital.

Unfortunately, we now discovered that there was not enough air pressure left in the starting air-bottles to start the port generator. Normally, this would not have posed much of a problem, the air-bottles could have been pumped up using the hand-start emergency diesel-driven compressor. Unfortunately, an inspection of this unit showed it to be in very poor condition. Frank and I examined it closely: there was a hole in the sump, the cooling pump was seized and the compressor valves were solid with rust. It really was in a bit of a mess. Apparently, this vital piece of equipment had not been used for some time. I was not very confident about our chances of fixing it without major replacement of parts – and goodness knows where we would get hold of them. Frank, however, reckoned he would be able to get it going. He had looked round the engineer's store and thought there were enough bits and pieces lying about to enable him to improvise some sort of emergency repair.

Based on my previous experience of Frank I wasn't too confident about the end result.

At about this stage Monty realised that he had left the spare batteries for his C.B. radio back at the Met. Station. The salvage team were now all fully occupied so I volunteered to go back for them. We certainly did not want to lose contact with our friendly Oudtshoorn undertaker; he was our only contact with the world. The Zodiac would not be available to pick me up at the lagoon, but by timing my journey correctly I would be able to wade across at low water.

I left the casualty just after noon on 27 April and reached base camp about 4.00p.m. Without the weight of the Honda pump on my shoulder it was a most pleasant journey and I was able to study the flora and fauna of the island. Along the shore there were plenty of Frank's hermit-crab friends, but no signs of his missing toothpaste – no gleaming white crab claws to be seen. The water depth at the lagoon was only about two feet, with some deeper pools here and there in which fish had been trapped by the ebbing tide. In one pool I nearly stepped on a four-foot-long sand shark; it shot away in a flurry of churned up sand and water, giving me the fright of my life. Much to my surprise, there were quite a few goats wandering about. Presumably, they had been introduced from Reunion, possibly to provide a source of meat and milk for the French garrison in the days before the airlifting of supplies became common-place. Mind you, when you consider the inhospitable nature of the island, it was a rather optimistic project. What puzzled me most about the goats was where they got water. The South Africans, based on their knowledge of the desert-like regions in Namibia and the Kalahari, reckoned that the only source would be the early-morning dew, which they would lick from the leaves of the bushes and trees, augmented by the odd overnight rain shower.

179

After spending a pleasant evening with the French weathermen I returned to the scene of action next day, again timing my journey to cross the lagoon at low water. No sand sharks to alarm me this time.

Back at *Kwang Myung 156*, I handed over the radio batteries to Monty and then went down into the engine-room to see how Frank was getting on with the compressor. On entering I heard the sound of running machinery. To my astonishment the sound came from the emergency compressor, which was running sweet as a nut, pumping up the air-bottles. Frank had worked non-stop for the best part of two and a half days on the compressor and, against all the odds, and my predictions of doom, had turned this rather derelict piece of machinery into a going concern. It was a really first class effort and the prospects of successfully refloating the stranded vessel had substantially improved. My previous views on Frank had been proved totally wrong and Monty's decision to pick him for the team a wise one. It just goes to show how wrong you can be about people.

A few months later, if I may digress for a moment, while bringing our office Christmas party guest-list up to date, my thoughts strayed back to Europa Island, and I added Frank McLoughlin to the list of guests. Frank duly attended, enjoyed himself immensely and kept us all entertained. At one stage someone nodded his head at the crowd round Frank and said to me, 'That guy's got a lot to say for himself, hasn't he.'

I smiled and said, 'Believe me, he can do a lot more than just talk.'

There was now enough air pressure in the bottles to start the port generator at 2.30p.m. The Honda pump had been connected into the cooling system and, with the bilge and general-service pumps operating on the bilge system, the water level in the engine-room was kept under control and no flooding occurred during the high water period. This was a major breakthrough.

Next day, 29 April, the SADS team were busy checking fuel lines with a view to filling up the main-engine daily-service tank. Frank was checking the main-engine to make sure it would be ready when needed, as well as keeping a watchful eye on the port generator. The generator operated all day, except at low water when the cooling water source dried up, and the engine-room bilge level was kept under control throughout the high tide periods. With power available the main air compressor was brought into service and the starting air-bottles were now fully charged.

During inspections on 28-29 April, the contents of the forward domestic reefer chamber were found to be starting to rot and work now commenced on dumping the contents over the side. The fish round the periphery of the fish hold had started to thaw out, but in the middle they were still frozen solid.

Both anchors, which had been lying on the ground together with several lengths of cable, were heaved up into the hawse pipes and secured. They would be required later to assist with refloating.

On 30 April dumping of the reefer chamber contents was completed and, with power available on the cargo winch, discharge of the fish hold commenced and, together with work on various other parts of the vessel, continued, despite the intervention of the local mosquitoes, until 9.00p.m. At this time low water conditions caused us to shut down the generator.

During the discharge of the fish cargo I initially assisted by operating the winch bringing the fish up out of the hold, while Monty and one of the divers heaved them over the side. After a few hours I found that the business of co-ordinating the movement of the winch lifting-hook with the signals of the man in the fish hold both difficult and tiring,

181

so I swapped jobs with the diver who was helping Monty to throw the fish over the side. This required more brawn than brain but I didn't find it nearly as tiring as operating the winch. I am sure there must be a moral there somewhere.

At high water in the afternoon the bow was observed moving gently in a long, low swell. The situation was improving.

Causeway Salvor arrived off the island at 1.30a.m. on 1 May and moved close to the casualty shortly after first light. Later in the morning two ten-inch polypropylene lines were passed ashore and made fast to bollards at the stern of the casualty.

Discharge of the fish hold continued for most of the day, while power was available. In the engine-room Frank was busy with modifications to the bilge system, in order improve the de-watering capability and make sure there was no chance of flooding during high tide. Approximately 1,000 litres of diesel fuel was transferred from the double bottom tank to the main-engine daily-service tank.

With the approach of high water the tug slowly took the strain on the two polypropylene lines and by 5.00p.m. was pulling on full power. There was a bit of movement at the fore end on the top of the tide, but no movement towards deep water. By 9.00p.m., on the falling tide, the casualty was firmly back on the bottom and the tug was requested to ease down on power. Slight tension was maintained on the towing connection through the night.

On the rising tide, at 5.00a.m. next morning, the tug increased power. At the same time, the main-engine was started and operated at 'full astern' to assist the tug. By 7.00a.m. there was no sign of any movement towards deep water and, with the tide dropping, both the main-engine and tug power were shut down.

Discharge of the fish hold continued all day while power was available. In this we were now assisted by two seamen from the tug, who had been ferried over in the Zodiac during the previous evening.

Following the failure of the attempts to tow the casualty, stern first, clear of the bottom, it was now decided to try and swing the bow round – during both the previous attempts there had been some movement at the fore end – and tow the casualty off, bow first. In this connection, the port anchor was slipped, the cable lowered to the ground and paid out aft. One of the ten-inch lines was then disconnected from the stern and made fast to the port anchor cable. However, when the tug started pulling the tow-line parted. Attempts were then made to pass another line but, during this operation, a rope became wrapped around the tug's propeller. By 5.00p.m. the tug had cleared the rope from the propeller and was back in position, ready to resume work.

While the tug was busy with its propeller problem, the salvage team had disconnected the remaining tow-line from the stern, dragged it forward along the starboard side and made it fast to the bollards and cable-stopper on the starboard side of the forecastle. This line was then passed to the tug and made fast. At 5.30p.m. the tug started to increase power. At 6.00p.m., with the tug towing on full power, the bow slowly began to swing to starboard. Over the next 45 minutes the casualty heading changed from 340 degrees to 150 degrees. There was some slight movement towards deep water but, by 7.30p.m. the dropping tide forced the effort to be abandoned until the next tide.

The stranded vessel was now lying at right-angles to the shoreline, bow to seaward, having been turned through 180 degrees, with a starboard list of about two or three degrees. In order to correct the list and trim by the head, the port forward reefer chamber was flooded. It was hoped that, in addition to bringing the vessel upright, this would reduce the ground reaction at the stern and aid the refloating effort on the next tide.

The tug started towing at 5.00a.m. Unfortunately, the main-engine could not be used to assist owing to an increase in the bilge level. The bottom damage under the engine-room must have been aggravated when the casualty was swung round, causing the ingress of water to increase; there had certainly been quite a bit of crunching and grinding as the vessel was pulled round. To cope with this increased leakage an additional four-inch salvage pump with associated suction and discharge hoses was transferred from the tug and rigged in the engine-room. While this was going on the tug eased down on power, just keeping enough tension on the tow-line to prevent the casualty moving back up the beach on the flood tide. In an attempt to operate the generator continuously, the Honda pump was rigged into the cooling system, so that water could be circulated from the flooded reefer chamber through the generator intercoolers and then back to the reefer chamber via the wash deck line.

This work was all completed by about 5.00p.m. which was just before high water. The tug came back up to full power and, as the four-inch pump was keeping the bilge level under control, the main-engine was started up on 'full ahead'. At 7.00p.m. the bollard on the forecastle collapsed but the tow-line remained secured round the cable-stopper. At 7.10p.m. the casualty started to move ahead and, shortly afterwards, was afloat in deep water.

Unfortunately, the movement over the ground as the vessel was pulled towards the deep water must have again increased the bottom damage under the engine-room, which now began to flood quite rapidly. The Honda pump, the four-inch salvage pump and the fuel transfer pump were hooked up on the bilge system to assist the bilge and

184

general-service pumps, and the flooding was contained. However, as a further precaution, three more salvage pumps were sent over from the tug.

The rest of the night was passed peacefully as the tug, with the casualty in tow, steamed slowly round the north of the island, awaiting daylight before anchoring off the Met. Station.

At 9.00a.m. on 4 May the tug commenced shortening the tow-line while approaching the anchorage and by 10.15a.m. had anchored, with the casualty secured alongside. The rate of water ingress into the engine-room appeared to be increasing and the additional salvage pumps sent over from the tug during the night had to be brought into use. By utilising all our available pumping capacity the water level in the engine-room was kept low enough for us to run the port generator. There was no shortage of cooling water now.

For the next two days the divers were fully occupied, sealing the holes and splits in the engine-room bottom plating, using a mixture of wooden wedges, lead and epoxy. Gradually, the rate of ingress was reduced and, one by one, we were able to shut down the salvage pumps until, by the evening of 5 May, we were able to control what little ingress remained with occasional use of the bilge pump.

SADS contract stipulated that the casualty was to be re-delivered to the owners, safely afloat in a safe port. We now considered the vessel to be safely afloat. The most convenient safe port for redelivery was Durban, where SADS would be returning in any event and where the vessel could be dry-docked for permanent repairs. While there were several ports closer than Durban, none had really adequate repair facilities.

By the morning of 6 May *Kwang Myung 156* had been secured for the tow to Durban. Leakage into the engine-room was minimal but it was considered prudent to retain two of the salvage pumps in the engine-room, rigged ready for immediate use in the event of an emergency. Frank and two of the divers remained on board as the riding crew.

With all preparations completed, *Causeway Salvor*, with *Kwang Myung 156* in tow, departed Europa Island for Durban at 4.00p.m. The passage was uneventful and *Kwang Myung 156* was safely berthed alongside the repair wharf at Durban on the morning of 11 May, ready for re-delivery to owners.

The rest of us had to wait until the afternoon of 7 May before flying back to Durban. I spent the morning checking out the southern part of the island. From the airstrip I walked round the lower end of the lagoon, then across to the south-eastern corner of the island and up the east coast to where we had worked on the Korean vessel. The only signs of life were the usual hermit-crabs scurrying about on the shore and a few goats wandering about in the bush, nibbling at leaves and the odd tuft of dried-up grass. At one spot I came across the old wreck of a jeep. What it was doing there I have no idea. Perhaps it had been part of the French geographical expedition; on the other hand, it may have been a French Army jeep. Whatever its origins it looked quite mysterious sitting there in the last place in the world you would expect to find a jeep – or at least the remains of one.

Along the three-mile stretch of coast south of where our Korean long liner had stranded I came across the remains of three shipwrecks. One had obviously been there for many years; it had been a wooden sailing vessel. Only the keel and some of the hull frames remained, sticking up out of the sand like the skeleton of some prehistoric beast. The other two were, by comparison, fairly recent. They were easily identifiable as Far Eastern fishing vessels, either Korean or Taiwanese. Both had broken up and looked as if they had been abandoned for at least five years, possibly more. It seemed strange that four vessels, if we include *Kwang Myung 156*, had all stranded along a three-mile stretch of the island coast.

While I was exploring the southern end of the island, Mike Smith, the head diver, visited the French Army camp, where the medical orderly offered to give him something for the nasty sores he had collected on his legs and feet during our adventures, and which had now turned septic. Unfortunately, the orderly miscalculated the strength of the antibiotic injection that he administered. The result was that Mike passed out and was being carried back to the Met. Station just as I returned from my exploration. I had been slightly bothered by a couple of sores on my legs where they had been cut on the coral and then gone septic. My usual field treatment of applying mercurochrome, which is the South African cure all for cuts and sores, had had no effect. I had also thought about going along to the army camp for treatment, but had decided, now that the job was over, to wait and go and see my local G.P. back in Durban. One look at the unconscious Mike convinced me that I had chosen wisely. However, Mike was a fit young man – he soon recovered from the army treatment and was back on his feet in a couple of hours.

During our various trips around the island in the Zodiac we had observed several large turtles swimming serenely just below the surface. None of us had sufficient knowledge to identify correctly which particular species of turtle these were. Although it was difficult to judge just how big they were, they did not look big enough to be the large Leatherback, which can be 2 metres long and weigh up to 600 kg (over half a ton). The ones we saw appeared to be about a metre in length and about 100 kg in weight. Most male turtles remain at sea all their lives and the females only come ashore to lay their eggs. What a pity we did not have a marine biologist with us, because the beach next to the

Met. Station turned out to be an important turtle-nesting site. Not only that, but on the afternoon of 7 May, just after I returned from my 'walk-about' round the southern part of the island, and while Mike was recovering from his antibiotic injection, turtle eggs buried in their nests under the sand started to hatch out. Monty and I were sitting on the beach enjoying a cold beer when we became aware of a disturbance in the sand a few feet away. Then, as we watched in amazement, a small dark-coloured creature, about four-inches long, emerged into the sunlight and slowly but purposefully started to head down the beach towards the water. It was a baby turtle. The rest of the salvage crew were quickly summoned to see this fascinating event. Our little turtle was followed by a dozen or so of his brothers and sisters, all of them making their way steadily down to the water. As we watched more baby turtles started to emerge from several other nests along the beach. To say that we were absolutely enthralled by this incredible spectacle would be a serious understatement. Cameras were grabbed and we were soon busy taking photographs of the little creatures as they slowly propelled themselves over the sand. They looked for all the world like little, battery-operated rubber toys.

The transit from nest to sea was not without danger – hungry predators in the form of frigate birds were soon circling overhead, alerted by the prospect of an easy and tasty meal. The frigate is a large sea bird-of-prey, found on most tropical coasts. Unable to land on the water, they either fish on the surface, or eat eggs, chicks or, as in this case, young turtles.

A frigate bird suddenly swooped down, and one of the little turtles was gone. Amid shouts of anger cameras were dropped, stones were grabbed and a hail of missiles directed at any frigate bird who looked as if he was getting too close to our turtles. The salvage crew patrolled up and down the beach throwing stones, or other suitable objects, at any frigate bird that came within range. Baby turtles that appeared to be making very slow progress towards the sea, thereby presenting the birds with an easy target, were carefully picked up, carried down to the edge of the beach and placed gently in the water. By the time the sun dipped below the horizon the migration of the tiny turtles from eggs to sea had been completed with a minimum of loss. Monty and the rest of us, feeling quite pleased at having thwarted the evil forces of nature, then retired to the Met. Station for a well-deserved beer or two.

Next morning, after our aircraft arrived from South Africa, we loaded the remaining items of salvage equipment, together with our personal kit, onto the plane, said goodbye to our friends at the Met. Station and the army camp, then took off for Durban. The job was over, apart from writing the report, of course – and that, as every surveyor will tell you, is the really nasty part.

SALVAGE LOG

Kwang Myung 156   Aground at Europa Island

26 April 1980

Casualty heading 010 degrees
With no power available, engine room remained tidal during the night of 25–26 April.

08h00   Work continued rigging salvage equipment on board casualty and moving remaining equipment from Met. Station.

12h00   Commenced overhaul of emergency air compressor.

Commenced overhaul of port generator cooling pump.

27 April 1980
Casualty heading 010 degrees

08h00   Engine room remains tidal.

Work continues all day on emergency air compressor and port generator cooling pump.

Unsuccessful attempt made to rig a hand-operated air compressor into the starting air system.

28 April 1980
Casualty heading 010 degrees

03h00   Completed overhaul of emergency air compressor and commenced pumping up starting air bottles.

10h00   Portable salvage pump rigged on board.

Port generator cooling pump boxed up.

14h30   Sufficient pressure in starting air bottles to start port generator but trouble again experienced with cooling system and generator shut down while salvage pump connected to cooling system.

22h00   Port generator restarted with cooling water supply from salvage pump.

Bilge and general service pumps running on bilge system and water level in engine room contained. No flooding during high tide period.

Main air compressor started and air bottles pumped up to working pressure.

29 April 1980
Casualty heading 010 degrees

06h00   Checking fuel lines in preparation for filling main engine daily service tank.

Port generator in use with bilge and general service pumps on bilge system and engine room maintained dry throughout high water period.

07h00   General service pump failed. Pump opened, found choked with coral.

Pump cleared tested and found satisfactory.

08h00   Commenced discharge of forward domestic reefer chamber.

10h00   Both anchors heaved in and secured in hawse pipes.

30 April 1980
Casualty heading 010 degrees

02h00   Port generator on load.

Evaporator pump running on cooling system and bilges.

04h00   Main engine daily service tank filled (1,000 ltrs) from fuel oil double bottom tank.

09h00   Generator shut down due loss of cooling water at low tide.

14h00   Generator back on load on rising tide.

Completed discharge of domestic reefer chamber.

Commenced discharge of fish hold.

18h00   Suspended fish hold discharge due mosquito problem.

21h00   Generator shut down due loss of cooling water at low tide.

01 May 1980
Casualty heading 010 degrees

01h30   Tug *Causeway Salvor* in sight.

03h00   Port generator on load.

08h30   Recommenced discharge of fish hold.

09h00   Commenced discharge of ballast from no. 4 double bottom tank.

10h00   Two 10 inch polyprop. lines passed from tug and made fast to stern bollards.

17h00   Tug commenced towing.

17h05   Port side bollard carried away. Tug continues towing.

18h00   Considerable movement observed at fore end.

No movement towards deep water.

21h00   Vessel sitting firmly on bottom on falling tide.

Tug requested to stop towing.

Two seamen transferred from tug to assist salvage crew.

22h30   Abandoned all work on deck due mosquito problem.

2 May 1980
Casualty heading 340 degrees

05h00   Tug resumed towing. Some movement observed at fore end.

06h00   Main engine started running full astern.

06h15   Turbo-charger overheating. Main engine stopped.

06h30   Cooling system adjusted and main engine restarted on full astern.

07h00   No signs of movement. Main engine stopped and tug requested to stop towing

Recommenced discharge of fish hold.

09h30   Port and starboard anchors slipped and secured in hawse pipes.

Port cable lowered to ground and paid out aft.

Portside towing line disconnected from stern and made fast to port anchor cable.

193

12h00	Tug commenced towing.
12h30	Port anchor cable parted.
15h00	Tug propeller fouled while recovering port tow line.
	Tug crew working to clear propeller.
	Starboard tow line disconnected from stern bollard and carried forward to forecastle and secured to starboard side bollard and cable stopper.
17h00	Tug reports propeller clear and ready to resume towing.
17h30	Tug commenced towing gradually working up to full power.
18h00	Heading 340 degrees. Bow slowly swinging to starboard.
18h10	Heading 010 degrees.
18h15	Heading 050 degrees.
18h20	Heading 120 degrees.
18h45	Heading 150 degrees
19h00	Very slight forward movement of casualty observed.
	Tide dropping and tug requested to stop towing.
19h30	Recommenced discharge of fish hold.
20h00	Commenced ballasting forward port reefer chamber to correct list and increase forward trim.
	Discharge of fish hold suspended due mosquito problem.
22h00	Forward port reefer chamber full.

23h00   Generator shut down due loss of cooling water at low tide.

3 May 1980
Casualty heading 150 degrees
Casualty now lying at 90 degrees to the shore line with bow to seaward.

03h00   Cooling water system restored on rising tide.

       Generator started and power restored.

03h30   Recommenced discharge of fish hold.

05h00   Tug towing on full power.

       No movement towards deep water.

       Unable use main engine due high water level in engine room probably caused by further damage to bottom when casualty was pulled round through 180 degrees during refloating attempt yesterday.

       Bilge pumps unable control ingress of water on rising tide.

07h00   Tug easing down on power.

09h00   One 4 inch salvage pump transferred from tug together with associated hoses.

10h00   Additional pump and hoses rigged in engine room.

10h30   Continuing discharge of fish hold.

11h00   Power maintained through low water period by using the Honda pump to circulate ballast water from the forward reefer chamber through the generator cooling system and then return to the reefer chamber via the wash deck line.

195

16h00	Water ingress into engine room under control.
16h45	Discharge of fish hold stopped.
17h00	Tug commenced towing and main engine running on full ahead.
19h10	Starboard aft bollard in way towing connection collapsed but tow line still secure.
	Casualty moving ahead away from the shore.
19h15	Casualty afloat and moving towards deep water.
19h20	Increased water ingress in engine room. Both salvage pumps in use but water level slowly increasing.
21h00	One 4 inch and two 3 inch salvage pumps transferred from tug together with necessary hoses.
21h30	Additional pumps rigged in engine room.
22h00	Leakage into engine room under control.
	Casualty under tow awaiting daylight to approach Met. Station anchorage.

4 May 1980

04h30	4 inch salvage pump in engine room failed.
05h30	2 inch salvage pump in engine room failed.
08h00	Bilge level being maintained by the 3 inch salvage pumps, the bilge pump and the main engine cooling pump operating on the bilge direct injection.
	Commenced shortening tow line for approach to anchorage.
09h45	Casualty secured to port side of tug.

10h15 Tug anchored off Met. Station with casualty secured alongside.

11h00 Divers commenced sealing holes and splits in way engine room bottom using wooden wedges, lead and epoxy.

Work continued throughout the day sealing the underwater damage.

5 May 1980

08h00 Work continued throughout the day sealing the underwater damage and preparing the casualty for the tow to Durban.

6 May 1980

10h00 Divers completed temporary repairs to bottom damage.

12h00 Preparations for tow completed.

16h00 Tug and casualty departed Europa Island for Durban.

11 May 1980

08h00 Tug and casualty arrived off Durban.

10h00 Casualty secured, port side to the quay, at the repair wharf Durban ready for redelivery to Owners.

Salvage agreement terminated.

Salvage Equipment Used

Tug *Causeway Salvor* 3,500 BHP
                        30 tonnes bollard pull

Pumps   2 × 3 inch Spate pumps plus hoses
                        1 × 4 inch Homelite pump plus hoses
                        1 × 2 inch Honda pump plus hoses
                        2 × 3 inch Hatz pumps plus hoses

Boats   1 × Zodiac inflatable dinghy and
                        outboard motor

Other Equipment   1 × Portable burning/welding set.

Surveyor's Comments

While this was not a major salvage operation, the salvors were handicapped by the very remote location of the casualty, and the very difficult nature of the terrain between the island air strip and the site of the stranding. Some idea of the problems encountered can be gauged from the fact that it took six very fit men almost three days to move a two-inch portable pump a distance of approximately five kilometres. In addition work on deck after dark proved to be very difficult due to the presence of hordes of very aggressive mosquitoes. Nevertheless, work was carried out after dark on all but two occasions.

In our opinion the salvors worked well in difficult and unpleasant conditions.

10

Nicaragua

From 1981 until 1985 I served with the Salvage Association in North America, firstly at Halifax, Nova Scotia, and then New York. The New York office was the head-quarters of the Salvage Association's operations in the Americas and co-ordinated all surveys throughout North, Central and South America and the islands of the Caribbean. For more than three years I had a most interesting time, travelling to marine casualties at exotic locations as distant as Punta Arenas, at the southern end of Chile, or McKinlay Bay, way up beyond the Arctic Circle in Canada's North West Territories.

On Monday 12 March 1984 I got back to my desk at the World Trade Centre in New York from a job in Barbados, only to be informed that the underwriters wanted someone to go down to Nicaragua ASAP. This new job was to investigate the cause, nature and extent of damage to a ship at the port of Corinto. The only information available was that the ship's name was *Los Caribes* and it was reported to have been damaged by some sort of underwater explosion. At this time Nicaragua was in the grip of considerable unrest, with right-wing opposition forces carrying out guerrilla warfare against the government. The New York Times carried almost daily reports of ambush, explosion, political assassination and general mayhem. It was, therefore, not surprising that there was no great rush of volunteers to go down to have a look at *Los Caribes*. So 14 March found me on a flight to Miami to connect with the Nicaraguan State Airline's flight to the capital city, Managua. On arrival there I was met by the local Lloyd's agent and briefed on the general situation in Nicaragua.

Violent opposition to a long history of widespread government corruption had brought the Marxist Sandinista guerrillas, led by Daniel Ortega, to power in1978. However, the Sandinista government's subsequent support of leftist guerrillas in neighbouring countries, mainly El Salvador and Honduras, caused the United States to give covert military aid to the anti-Sandinista Contra guerrillas from about 1980 onwards. In the past few weeks the Contra guerrillas had widened the scope of their activities by carrying out several attacks on the Pacific coast ports of Corinto and Puerto Sandino. These attacks had generally taken the form of fast, light craft entering the ports at night and directing machine-gun and rocket fire at port facilities.

The attack on the port of Corinto had taken place around midnight on 29 February – 1 March. Then, at around noon on 1 March, while the Dutch dredger *Geopotes VI* was proceeding inbound through the approach channel after dumping a load of spoil outside, an underwater explosion occurred, causing severe damage to the forward underwater hull of the dredger. Following that incident everything went quiet for the next six days; but on 7 March the freighter *Los Caribes* was damaged by another underwater explosion. In Managua the consensus of opinion was that the explosions were the result of the Contra guerrillas laying mines in the channel. The mining of harbours, however, seemed to me to be a bit sophisticated for the Contras, so if they were indeed responsible they must have had quite a bit of help, in the form of equipment and training, from an organisation experienced in such matters.

The morning after my arrival I set out by car for Corinto,

Map 13 Nicaragua and the neighbouring states of Central America

140km to the north-west. The journey took me along Nicaragua's Pacific coastal plain which extends in from the coast for about 60km before giving way to the central interior mountains which rise to a height of 2,438 metres, or nearly 8,000 feet. The coastal plain is, however, bisected by a ridge of volcanic mountains. This line of mountains runs in a north-west/south-east direction, more or less parallel with the coast and the main road from Managua to Corinto, and provides a very scenic backdrop to the drive. At the north-western end of the ridge, near the major road junction of Chinandega, is the highest of the four main mountains – the San Cristobal volcano, with its summit at 1,745 metres and a permanent plume of smoke rising into the blue, tropical sky.

There were numerous police, or army, check-points along the road where the boot of the car was opened and inspected before we were waved on. Everybody appeared very friendly. In fact, almost everyone I dealt with in Nicaragua was polite and friendly, especially in the hotel and the state airline. After an hour and a half we reached Chinandega where we turned left, then journeyed for another half an hour and through several more security check-points, to reach Corinto at about noon. *Los Caribes* was berthed alongside the port's container pier, but before I could gain access I had to obtain a security pass from the police. In this connection I was taken to the Harbour Police Station and ushered into the presence of the officer-in-charge to whom I explained the nature of my business. He was a rather surly-looking police captain who, when I arrived, appeared to be re-assembling an AK-47 rifle. He took no notice of me whatsoever, just carried on doing what he had been doing before I came into his office. About 15 minutes of strained silence followed, with me standing fidgeting in front of his desk while 'El Capitano' fiddled about, trying to get all the parts of the rifle back together in

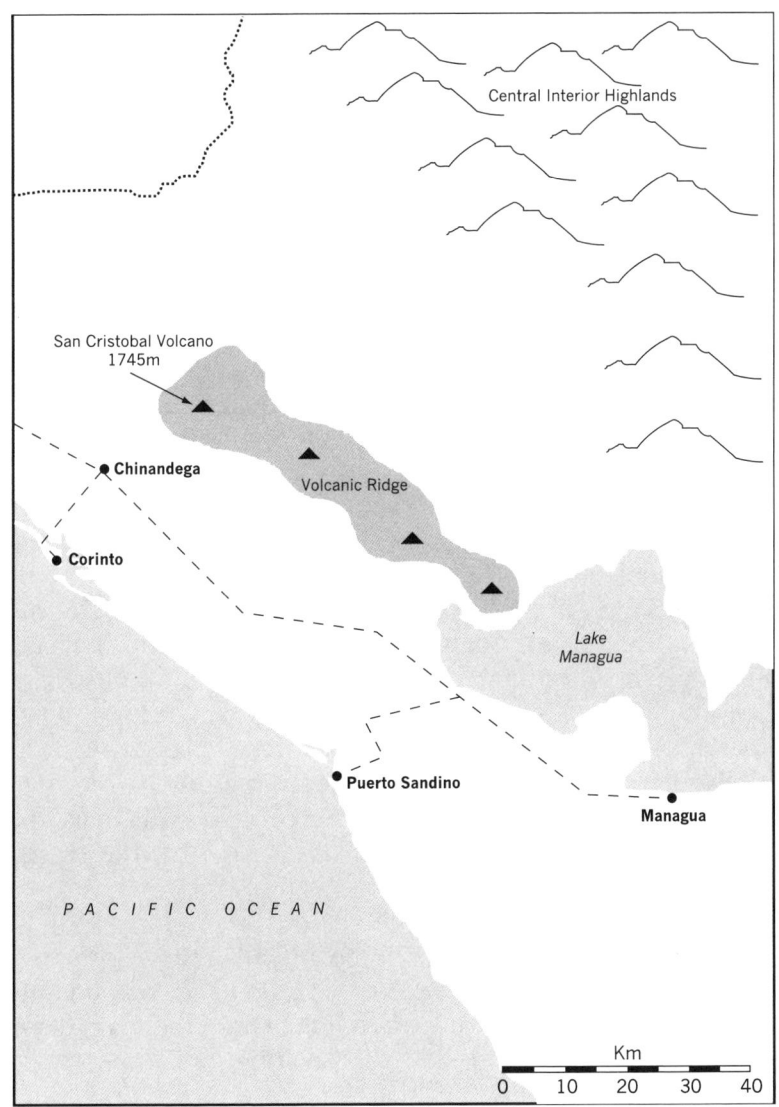

San Cristobal Volcano
1745m

Central Interior Highlands

Volcanic Ridge

Chinandega

Corinto

Lake
Managua

Puerto Sandino

Managua

PACIFIC OCEAN

Km

0   10   20   30   40

Map 14 Pacific coast of Nicaragua

the right place. Suddenly, without even looking up, he reached over for a pad and scribbled something on it, tore out the page and handed it to me. It was my permit to get onboard *Los Caribes*. 'Mucho gracias, Senor,' I said in my best survival Spanish and left his presence with as much haste as dignity would permit.

Los Caribes was a 5,018 gross ton, three-hold cargo vessel, built by the Burntisland Shipbuilding Co. Ltd., on the shores of the Firth of Forth, in 1969. The owners were a Caribbean multi-national shipping company based in Costa Rica. She had arrived at the anchorage about 6.00p.m. on 7 March on completion of a part-loaded voyage from Havana. A cargo berth was not immediately available but, as the vessel was short of water, she was allowed to enter port for the purpose of taking on fresh water. Having completed that operation at about 10.00p.m. the vessel undocked and, shortly afterwards, started to proceed back down the channel to the anchorage to await her turn for a cargo berth. At 11.10p.m., while in mid-channel in the vicinity of the No. 3 and No. 4 channel buoys, a heavy explosion occurred under the vessel causing extensive damage to both the hull structure and sundry deck fittings. After spending a nervous night at the anchorage, *Los Caribes* was allowed to dock under emergency conditions at the container pier early next morning, where an assessment of the damage was carried out and discharge of the cargo commenced.

By the time I arrived on board on the afternoon of 15 March the vessel was free of cargo and all areas were pretty well accessible for inspection. The damage was mainly located in way of the No. 2 cargo hold where the main deck, 'tween deck, starboard side shell-plating and the forward bulkhead were badly buckled and torn open at several locations. The side shell and bulkhead damage extended over the full depth of the hold, from bilge to main deck.

In addition, the No. 2 main and 'tween deck hatch covers and coamings were badly distorted and fractured, together with the cargo winch platform between the Nos. 1 and 2 hatches. Numerous deck fittings including the fire main, cargo derricks and electrical cable conduits, had also been damaged to varying degrees of severity.

Corinto had no ship-repair facilities, the nearest being at Balboa, at the Pacific end of the Panama Canal, or at the Dutch island of Curaçao in the Caribbean. My first task was, therefore, to decide how to patch up the damaged ship in order that she could proceed safely to either Balboa or Curaçao. Then, once I had organised the temporary repairs, I started work on a permanent repair specification that would cover all the work necessary to return the ship to its pre-explosion condition. This took several days to complete but, on 19 March, I was able to hand the completed specification to the owners' representative so that he could send it off to the shipyards at Balboa and Curacao and obtain quotations for carrying out permanent repairs.

Owing to the lack of adequate repair facilities at Corinto it took some time to complete the temporary repairs, but eventually the necessary work got done and *Los Caribes* was able to proceed under her own power, though at reduced speed, to Balbao where she was dry-docked for a more detailed examination. This revealed the damage to the hull structure to be more extensive than the afloat inspection at Corinto had indicated. Shipyard repair quotations based on the repair specification, plus the additional damage found on drydock at Balbao, were close to the insured value. The underwriters, taking into consideration the cost of the temporary repairs carried out at Corinto and various casualty-related expenses incurred by the owners, in addition to the shipyard repair quotations, eventually reached a settlement with the owners on the basis of a 'Constructive Total Loss'.

Meanwhile, the problem of mines at the Pacific coast ports continued into April, with a Russian tanker being damaged at Puerto Sandino on 21 March and, on 30 March, a Japanese freighter being damaged at Corinto. In total, twelve ships were damaged over a period of about six weeks.

The puzzle of how the Contra Guerrillas – and it seemed certain that they must be responsible – were able carry out such a sophisticated type of warfare remained to be solved. Several media reports hinted at the involvement of the CIA, but these were unsubstantiated. Then, on 10 April, the following press report was filed at Washington DC:-

'The Central Intelligence Agency has directed a covert operation for more than six weeks to lay mines in two Nicaraguan harbours on the Pacific coast and the work has been carried out in part by former Latin-American military personnel,' defense sources said. 'The acoustic mines have been sown in the ports of Corinto and Puerto Sandino under cover of darkness from speedboats operating from a freighter,' the sources said. 'Acting as a mother-ship for the storage and arming of the mines with up to four speedboats moored alongside, the freighter was at least 25 miles from the Nicaraguan coast, out of range of artillery,' the sources said. The sources, who requested anonymity, confirmed reports about the mining operation that have appeared in the media since Friday (April 6) but they added several details. 'The small acoustic mines, which are triggered by the sound of an approaching vessel, cause only minor damage and thus are designed in this case to convey more of a psychological than a military impact, with the object of keeping cargo ships from calling at Nicaraguan ports and thus crippling the Nicaraguan economy,' the sources said.

United Press International.

A further report dated 11 April stated that:-

In Washington DC, late yesterday, Congressional and defense sources said the CIA-directed ship that was supporting mine-laying in Nicaraguan harbours had left the area.

United Press International.

The mystery was solved!

The background to this affair was that in December 1981, President Reagan approved an initial amount of $19.95 million to support and conduct paramilitary operations against Nicaragua. However, in 1982, the United States Congress passed a law prohibiting US support for the Contras. President Reagan's National Security Council then set up a system to circumvent Congress and carry on supplying military aid to the Contras. Much of this aid took the form of Soviet-bloc weapons captured from the Palestine Liberation Organisation (PLO) by the Israelis and then purchased by the United States, using funds obtained from the secret sale of US weapons to Iran. The arms sales to Iran were an attempt to secure the release of American hostages held in the Middle East. With regard to the laying of mines at Corinto and Puerto Sandino, this appears to have been entirely a CIA operation, with the Contras told to take responsibility so as to cover up United States' involvement, for, of course, the mining of the harbours broke numerous national and international laws.

This whole clandestine can of worms eventually came to be known as the Iran-Contra Scandal.

Initially, I was a bit sceptical about the mines being designed to cause only minor damage. The damage to *Los Caribes* had been fairly severe and had resulted in the ship being declared a 'Constructive Total Loss'. However, had the CIA and the Contras used a more powerful weapon it would have blown a hole in the bottom and probably sunk the ship. More importantly, it might well have caused loss of life among the crew and that was something the CIA would have wanted to avoid. Bumping off Sandanista government personnel was one thing, but killing seafarers going about their lawful business would have been something quite different. So the UPI report about the mines being designed to cause only comparatively minor damage was probably correct.

One thing puzzled me at Corinto: it appeared almost certain that the mines were laid under cover of the rocket attack on the night of 29 February-1 March. The damage to the Dutch dredger at midday on 1 March was consistent with that scenario. Following the rocket attack and the damage to the dredger, the Sandanista security forces were keeping a sharp lookout along the sea approaches to Corinto and it was unlikely that any further mine-laying was carried out during the next couple of weeks. A review of ship movements at Corinto showed that after the damage to the dredger four or five ships entered and left the port without incident; then, on 7 March, *Los Caribes* was damaged. Did that mean that further mines were laid on 6-7 March. Bearing in mind that the Sandanista forces were patrolling the harbour and the approaches, that seems unlikely. Or was it that mines laid on 29 February-1 March had, for some reason, not been detonated by the ship movements in the channel during the period of 2 March to 6 March. But then why had one been detonated by *Los Caribes* on 7 March?

A contact in the Royal Canadian Navy at Halifax, Nova Scotia, supplied a possible answer to this puzzle. Apparently, acoustic mines used in shallow water, like the Corinto approach channel, often had a timer incorporated into the detonator. With the timer set on zero the mine would be detonated by the first ship-generated sound-wave to pass over it. On the other hand, if the timer was set at six, for example, then five ships could pass over the mine safely; the timer would click round only one notch as each sound-wave passed over the mine. As the sixth ship approached, the timer would now be at zero and the sound-wave from the ship would detonate the mine. The idea was to cause confusion by making the port security forces think that, in view of the uneventful passage of several ships, their harbour was free of mines when, in fact, a deadly danger still lurked beneath the water.

11

Newfoundland

While based at Halifax, Nova Scotia, in the early 1980s I attended several marine casualties on the adjacent island of Newfoundland. I found Newfoundland, and especially the inhabitants of the island, to be fascinating and quite unlike any other part of Canada. On my first visit I was examining some piece of damaged machinery in a workshop on the St John's waterfront when the workshop foreman asked somewhat conversationally, 'How long you bin over here then, bye?' I replied that I had been in Canada for only a couple of weeks. The foreman gave me a rather old-fashioned look, then said, 'Ah, but this ain't Canada, bye. This be Newfoundland.' I felt suitably rebuked.

The island of Newfoundland lies off the north-eastern coast of the North American land mass and is closer to Europe than any other part of the continent. The most easterly point, Cape Spear, is actually closer to Ireland than it is to Thunder Bay, Ontario; and it is, therefore, not surprising that this was the first part of the 'New World' to be explored by Europeans.

The original inhabitants of Newfoundland were the Beothuk, an Alkonkian-speaking race of hunters, who would appear at one time to have occupied most of the island. They gradually became extinct; the last known Beothuk died in St John's in 1829. The present day inhabitants of aboriginal ancestry are the Micmac, whose homeland included what is now Nova Scotia and Prince Edward Island.

Archaeological evidence indicates that Norse voyagers reached Newfoundland around 1,000 AD and established what is the earliest known European settlement at L'Anse-aux-Meadows on the northern tip of the Northern Peninsula, near the entrance to the Belle Isle Strait. The Norse, however, left no permanent legacy and it was not until John Cabot sailed from Bristol in 1497 that the next documented voyage from Europe took place. Europe was expanding, and part of this process, together with a growing interest in finding a direct western route to Asia, was the progressive exploration of the Atlantic Ocean. As far as can be ascertained, Cabot carried out a reconnaissance of what is now the Bonavista and Avalon Peninsulas. In the wake of Cabot's voyage of discovery, migratory fishing vessels from France, Portugal and Spain began to harvest cod off the coast of Newfoundland in the summer months. These vessels sailed from Europe in the spring, returning with their catch of salted codfish in the autumn. During the summer season the fishermen established temporary settlements around the Newfoundland coast. Later, some of them chose to 'winter over' in their temporary villages instead of returning home and, eventually, many of these temporary settlements became the established outport fishing villages of the present day.

By the seventeenth century the presence of Iberian fishing vessels in Newfoundland waters declined and they were gradually replaced by boats from the south-west counties of England. Eventually, these English boats from the West Country came to dominate the fishing industry. Then, from about the mid-eighteenth century onwards, came a progressive move to operate from permanent settlements on Newfoundland instead of the seasonal migratory system of the early years. The settlers came mostly from the English West

Country and the south-east of Ireland. The population mix they created remains almost unchanged today and, coupled with the geographical isolation of the island, has resulted in the very distinctive Newfoundland speech, almost completely devoid of any North American intonation.

The capital of Newfoundland is St John's, located on the eastern side of the Avalon Peninsula, almost twelve hundred miles further east than New York. It is built round a naturally sheltered deepwater harbour. John Cabot probably anchored there during his voyage in 1497 and some sort of permanent settlement was established in 1528, making it the oldest town in North America, where numerous ships were using the harbour 40 years before the *Mayflower* sailed from England.

Britain assumed sovereignty over the island in 1713, making Newfoundland the oldest part of the British Empire. It remained as a separate British Colony until 31 March 1949 when it entered confederation with Canada.

Most of the original settlers were involved with the fishing industry and made their homes in little villages, known as outports, strung along the rugged coastline – a coastline consisting of nearly six thousand miles of cliffs, rocks and numerous submerged reefs that Newfoundland seamen refer to as 'sunkers'. The name is most appropriate. Apart from the nature of the coast, there is the weather. From October through to May a series of gales drives eastwards down the valley of the St Lawrence to meet the Atlantic. Now and again these gales are supplemented by hurricanes which are spawned in the Caribbean, then drive north-east up the eastern seaboard as far as Labrador. In addition to the gales there are another two deadly enemies to deal with – ice and fog. Great continent-sized masses of ice are carried down with the Greenland current to the seas off Nova Scotia and Newfoundland and create major hazards to navigation. The second enemy, fog, is perhaps the most deadly of all.

It can lie, day after day, like a smothering grey blanket over sea and land and, together with the rock-girt coast and the ever present icebergs, is a destroyer of ships and men. Many of the names of the bays, inlets and headlands along the coast give an indication of its nature: Bay of Despair, Malignant Cove, Misery Point, Mistaken Point, Cape aux Morts, etc.

However, the combination of this extremely hostile maritime environment and the Newfoundlander's dependence on the sea for his livelihood has had one very positive outcome: it has produced what are arguably the finest seamen in the world.

To help them face the rigours imposed by working in the fishing industry and the very unfriendly climate – once described to me by a fisherman as being – 'nine months winter and three months bad weather' – the Newfoundlanders are wont to fortify themselves by consuming fairly large amounts of rum. Rum is to Newfoundland as malt whisky is to the Highlands of Scotland. In days gone by when rum supplies ran low the inventive Newfoundlanders produced their own version called 'Screech'. This drink is made by pouring boiling water into empty rum barrels to dissolve whatever rum residue remains. Molasses and yeast are then added and the resulting mixture allowed to ferment for some time. The effect of this drink on those unaccustomed to it can best be described as 'lethal'. These days, bottles labelled 'Screech' can be purchased in the local liquor stores. It is not, however, old-fashioned 'Screech' of days gone by. It is more like a rather poor quality rum. The real 'Screech' is obtainable only in the back rooms or kitchens of remote outport homes.

In April 1982 I received instructions from the Dominion Insurance Company in Toronto to investigate the condition of a fishing vessel named *Belle Isle Lass* that had apparently got herself into trouble at some place by the name of Parson's Pond in Newfoundland. A look at the map showed that Parson's Pond was located on the western side of Newfoundland, just to the north of the Gros Morne National Park. The nearest airport was at Deer Lake, about 70 miles to the south-east, so next day saw me on Eastern Provincial Airline's Halifax-to-Gander flight, which stopped at Deer Lake. At the airport I rented an Avis car and set off for Parson's Pond. The road took me along the scenic coast of the Gros Morne National Park, through Rocky Harbour, Sally Cove and St Pauls, to arrive at Parson's Pond at about 2.00p.m. in the afternoon. Despite its being April the weather was fairly cold, with the temperature hovering not far above zero. Although the roads were clear, snow lay deep on the ground and the white snow on the Long Range Mountains contrasted with the dark, leaden clouds which held the promise of more to come.

Parson's Pond, at first glance – and even at second glance – was not very inviting. About a dozen inshore fishing boats lay at their moorings. A small wharf, a couple of fish stores and two score or so of houses lay around the inlet in a pattern of various shades of grey. It was cold and damp, with sleet driving almost horizontally in front of a brisk nor'easter. Now where was the *Belle Isle Lass*? None of the fishing boats that I could see appeared to be in a distressed condition, so I stopped the car and asked a local if he could direct me to the damaged vessel, or its owner, a Mr Eli Hallohan. He replied with that old-fashioned courtesy so prevalent among the people of the outports.

'Well now, Skipper, I think ye'll foind the spiled boat round t'other side o' the harbour. Just ye be following t'road for half a mile or so and ye'll see them on t'shore.'

He was correct; half a mile on I spotted a boat lying on the shore, just clear of the water. Parking the car I changed

Map 15 Newfoundland

215

into my working gear, then walked down to the boat which was, indeed, *Belle Isle Lass*. A gentleman in the working rig of a fisherman appeared from the depths of the vessel and proved to be the owner and skipper, Eli Hallohan. After we had introduced ourselves he told me the story of how *Belle Isle Lass* had ended up in her present situation.

Eli was not from Parson's Pond but came from Port Saunders, some 50 miles to the north. About 14 days previously he had sailed, with his crew of two, from Port Saunders to hunt for seals on the pack-ice in the Belle Isle Strait and off the Labrador coast. After one week they had collected roughly 200 seal pelts, which pretty well filled the boat, and decided to proceed in to Parson's Pond to land the catch. Now why he chose to offload the seal pelts at Parson's Pond and not his home base of Port Saunders, with which harbour they were no doubt very familiar, I cannot now recall. Whatever the reason, the weather as they approached Parson's Pond was the usual Newfoundland sort of weather – in other words, pretty foul. A heavy swell was running up into the harbour and, in Eli's own words, 'She was comin' on to blow somethin' awful, sorr.' Anyway, the end result was that *Belle Isle Lass* ran onto a sand-bar near the harbour entrance and was swamped by the heavy seas. Apart from flooding the boat throughout, the seas washed away all the valuable seal pelts. A couple of days later, when the weather had abated somewhat, Eli, with the assistance of a couple of the local boats, managed to pump-out his stranded craft and refloat her. Then they towed her up the harbour and, with the assistance of a caterpillar tractor, hauled her up clear of the water to where she now lay.

Belle Isle lass was a fairly standard Cape Island class of inshore fishing vessel, with a wooden hull and raised fore-castle. The hull was constructed of spruce planks laid over laminated frames. The overall length was 40 feet and she

216

had a beam of 15 feet. From forward the interior of the hull was divided as follows: accommodation space (or cuddy) with sleeping berths, galley with oil stove, toilet, engine compartment, fish hold, and an aft storage-space or lazerette.

Main propulsion was by a 200 hp GM Detroit diesel engine, driving a stainless steel propeller shaft through a reverse reduction gearbox. A power take-off supplied the necessary power for the hydraulically operated fishing-winches. The wheelhouse was equipped with an SSB trans-ceiver, VHF radio, radar and sonar.

An external inspection showed that several of the hull planks had been torn off and numerous others split and holed. At the stern the rudder was twisted and hanging partly off; the propeller blades were all heavily distorted, making it highly likely that the propeller shaft was bent.

Internally, all of the wheelhouse, accommodation, galley and machinery fittings were badly water-damaged owing to the boat having completely flooded throughout after she grounded. In addition, the fuel tanks were flooded with salt water and the entire system contaminated.

After completing my inspection of the damaged craft, I did a quick estimate of the extent of the necessary repairs. Back in the office in Halifax we had numerous examples on file of similar damages to Cape Island type fishing vessels and I had picked out some salient facts and figures before I left for Newfoundland. A quick comparison of what I considered would be the necessary repairs, including moving the vessel to a suitable repair facility – probably Ste-phenville, about 125 miles to the south – with the repair cost information in my notebook, showed that the cost of repairs would greatly exceed the insured value of the vessel.

I told Eli of my conclusions regarding the repair of his boat and advised him that, as soon as I returned to Halifax, I would send off a telex to Dominion Insurance in Toronto, telling them that repairs could not be carried out within the insured value. The telex would then be followed by a formal survey report. Hopefully, that would ensure there would be no undue delay with the settlement of his claim.

Eli seemed to be reasonably happy with this but, of course, there was nothing I could do for him regarding the loss of his 200 or so valuable seal pelts. They were not covered by his insurance. I said to him, 'It's a pity though about losing all your seal pelts, Mr Hallohan, but I'm afraid there is nothing I can do for you on that score.'

'Don't you be worryin' 'bout that, sorr. I got more than a hunnerd and fifty o' them back, you know.'

'How on earth did that happen?' I asked him. 'I thought you said they were all washed away when the boat flooded after you grounded.'

'Ah, well now, sorr, y'see most o' them fetched up on t'shore o'er there and t'local people gathered them up. They knew they come from my boat so they just give me them back.'

I looked at Eli in silence, then gazed around me. It was a bleak-looking spot indeed – a grey, cold and grim place in the falling sleet. Parson's Pond, however, had a hidden beauty of its own. Its outward appearance was deceptive for it disguised the inherent goodness and kindliness of its population, qualities that outsiders like me have found to be endemic to the people of Newfoundland.

While working in New York, a year or so later, I sometimes told some of the hard-nosed insurance brokers and adjusters who worked in that great city the tale of how Eli got his seal pelts back after his shipwreck at Parson's Pond. But I am afraid that the rock-girt, stormy, foggy coast of Newfoundland and its rugged, but warm-hearted, people may just have been a step too far removed from the streets of New York for them to fully appreciate the story.

218

12

Chatham Island

During the afternoon of 31 December 1999 we sat round our television-set in rural Perthshire and watched the celebrations as Australia and New Zealand moved into the new millennium – or, at least, what was popularly referred to as the new millennium – twelve hours or so ahead of us. Personally, I considered that the new millennium didn't start until 1 January 2001, but that's another matter. Even before New Zealand started to celebrate we caught glimpses of the excitement on the Pacific Islands to the east. Tonga and Chatham Island lie just west of the International Date Line making them the first inhabited parts of our planet to enter the new millennium. We saw Polynesian dancers on Tonga and what must surely have been the very first wedding of the millennium on Chatham Island. As I watched the satellite link-up to those far-off lands, I wondered how many people watching in Scotland knew where Chatham Island was, or had even heard of it – probably not many. Indeed, despite having travelled round the world for more than 40 years, it was only in June 1993 that I 'discovered' this very remote island.

Chatham Island lies on latitude 44 degrees south, longitude 176 degrees west, 600 miles east of New Zealand's South Island, more or less abeam Christchurch. Beyond lies the empty Pacific with no landfall until the coast of Chile is reached, almost 7,000 miles to the east. The island lies in

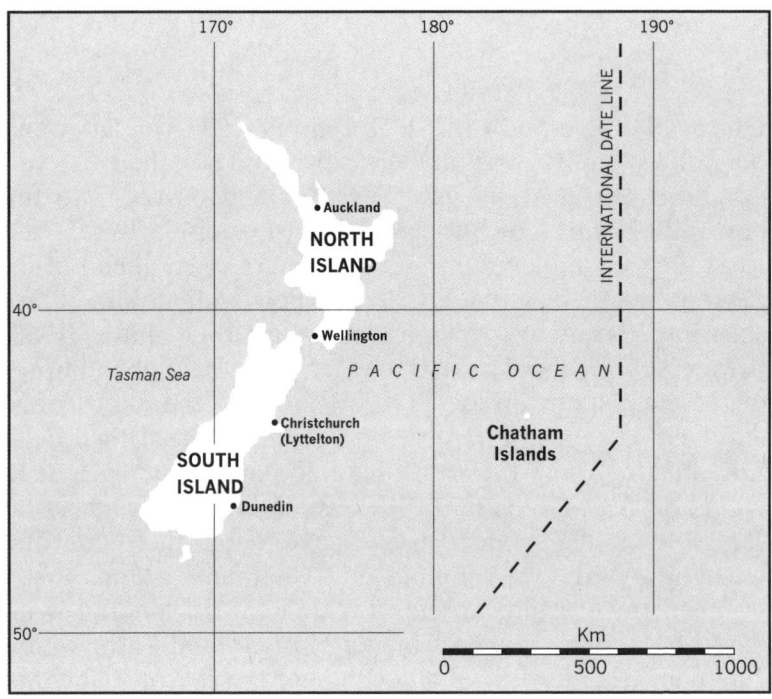

Map 16 New Zealand and the Chatham Islands

the path of the notorious 'Roaring Forties', that belt of almost continuous gales that sweeps round the globe bringing bleak weather during the Southern Hemisphere winter. It is part of a mini-archipelago, consisting of four principal islands – Chatham, Pitt, South East, and Mangere – and numerous other smaller rocks and reefs. The group, collectively known as 'The Chathams', marks the point where an elongated submarine ridge, the Chatham Rise, breaks the surface of the Pacific. This submarine range extends eastwards from the Banks Peninsula on the eastern side of New Zealand's South Island to a point about 100 miles to the east of The Chathams. Only the two main islands, Chatham and Pitt, are inhabited. In 1991 the population on Chatham was in the region of 700, with about 50 people on the much smaller Pitt Island, which lies some 12-miles southeast of its larger neighbour.

The main feature of Chatham Island is the Te Whanga Lagoon, which almost cuts the island in two. Originally the lagoon was a deep inlet on the east coast which, in time, became enclosed by a line of sand dunes on the seaward side. The first air service to serve the island started in 1942, using Sunderland flying-boats that landed and took off in the sheltered water of the lagoon. The landscape consists of low, rolling hills with the highest points being volcanic rock. Originally, the island was covered in bush but now large areas have been cleared for grazing, although a substantial part remains covered with small trees, shrubs and regenerating bush. The coastline is made up of beautiful, sandy beaches and cliffs of volcanic rock. Located on the west side of the island at Petre Bay is the main settlement of Waitangi, where there is a wharf which provides a secure berth in all but the roughest weather.

The original inhabitants of the island were the Moriori, a Polynesian people, who would appear to have sailed in their canoes from New Zealand. Tools and artifacts identifiable

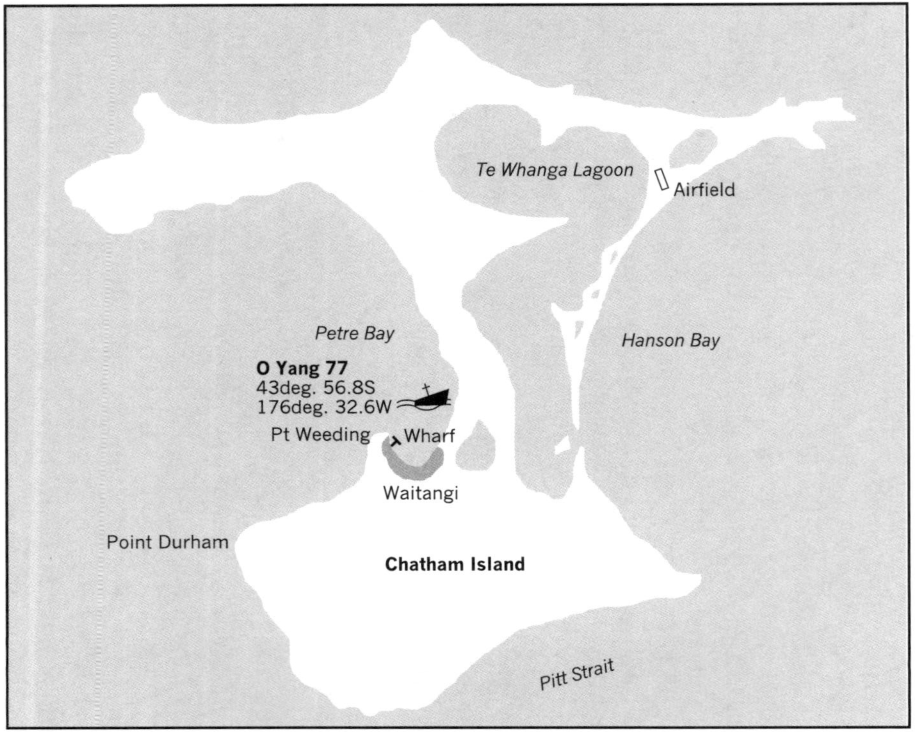

Map 17 Chatham Island

Airfield

Hanson Bay

Te Whanga Lagoon

Pitt Strait

Chatham Island

Waitangi

Wharf

Petre Bay

O Yang 77
43deg. 56.8S
176deg. 32.6W

Pt Weeding

Point Durham

222

with New Zealand have been found in their ancient living sites. Their ancestors were probably also the ancestors of the New Zealand Maori. Nobody is really sure when the Moriori first arrived. Some experts believe it could have been as early as 900AD. Others consider it more likely that they arrived about the 15th century. The Moriori named the island, Rekohu, meaning 'misty skies', after the characteristic weather conditions of fog and mist, produced when the cold currents from Antartica meet the warm streams from the north.

The first Europeans landed on the island in November 1791. The British brig HMS *Chatham*, under the command of Lieutenant William Robert Broughton RN, was on passage to Tahiti in company with HMS *Discovery*. The vessels became separated during storms to the south of New Zealand. HMS *Chatham* was set well to the east of the intended track and on 29 November sighted an unknown island. The brig anchored off the north-eastern point of what was, in fact, Rekohu, and Lieutenant Broughton, with eight of his crew, went ashore in the ship's cutter and claimed the island for King George III. Unfortunately, while attempting to barter with the islanders and obtain fresh water a scuffle broke out and one of the islanders was killed. The first Moriori contact with Europeans had not been auspicious. Broughton named his discovery, Chatham Island, and the spot where he and his men landed, Skirmish Bay.

Following the departure of HMS *Chatham* in 1791, the next Europeans to visit the island were seal hunters from Australia and New Zealand. They, in turn, were followed by American and European whalers. Permanent European settlers began living on Chatham Island about the late 1820s. In 1835 a group of 900 Maori arrived from New Zealand and, after killing about 300 of the original Moriori, established their right of ownership by conquest. Lieutenant Broughton's little skirmish in 1791 was pretty small beer by comparison. Major European settlement of The Chathams began in the 1840s when traders, farmers and missionaries arrived. The islands became part of New Zealand in 1842 and are presently controlled by the Chatham Islands Council. They also have a Member of Parliament, whom they share with the electorate of Wellington Central.

Over the years, various industries have flourished and died on Chatham Island. First there was seal-hunting, then whaling, sheep-rearing and, in the 1960s and 1970s, crayfishing. By 1974 the crayfish had been almost wiped out and today the main industries are fishing, sheep and cattle farming, and tourism, which has been given a big boost by the millennium celebrations. Many New Zealanders and Australians, and others from much further afield, travelled to Chatham Island to be among the first people on the planet to enter the new millennium.

My involvement with Chatham Island started on 16 June 1993 when we received an urgent message in Singapore from Ankuk Fire & Marine Insurance Co. Ltd., in Seoul, advising that one of their client's vessels had run aground at Chatham, near Christchurch in New Zealand. The message contained the usual instructions for us to get a surveyor to the vessel ASAP and find out what could be done to refloat it. A careful check of the Canterbury area of New Zealand's South Island failed to show any signs of a Chatham near Christchurch, but a quick phone call to our representative in Auckland revealed that the Korean vessel was actually aground at Chatham Island, 600 miles east of Christchurch. The owners of the grounded vessel, the stern trawler *O Yang 77*, port of registry Pusan, had an agent in Christchurch and he had already been in touch with our man in Auckland. Following a three-cornered discussion involving me in Singapore, our rep. in Auckland and the owners' rep. in Christchurch, the owners entered into a contract with Southern Tug & Barge, Wellington, for them to refloat the trawler and redeliver her to the owners in a suitable New Zealand port. The contract commenced at 9.10p.m. on 15 June and, almost immediately, the tug *Southern Alpha* departed Wellington with an ETA at Chatham Island of 2.00a.m. on 18 June.

Initially, we had tried to engage Southern Tug & Barge on a Lump Sum No Cure – No Pay contract, but without success. Local contractors able to offer salvage services around New Zealand are generally not geared up for difficult salvage operations and financial liquidity limitations restrict their ability to enter into No Cure – No Pay contracts. In this particular case, the remote geographical location and the unpredictable weather conditions around Chatham Island made the contractors even more unwilling to quote a lump sum to refloat the trawler. We were, therefore, forced to engage Southern Tug & Barge on a daily-rate basis. This is never the best type of salvage contract, unless the job is fairly straightforward and can be completed quickly, for, should difficulties be encountered and the operation become protracted, costs can spiral out of control. Given the remoteness of The Chathams and the potential for logistical problems should the refloating prove too difficult, there was a strong possibility that this could be an expensive operation.

O Yang 77 had been calling at Waitangi for the purpose of picking up four Indonesian seamen who had flown in via New Zealand on 15 June, when the stranding occurred. The approach to the island was made from the south. After passing Pitt Island the vessel proceeded north-east towards Point Durham at the southern end of Petre Bay. The master, who had advised that his ETA was 5.00p.m. on 15 June, was instructed by the local fisheries surveillance officer to approach Waitangi and stand off two cables from the wharf while maintaining a listening watch on Channel 16. The weather at the time was: wind westerly, 35 knots; swell

3-4 metres. After rounding Point Durham shortly after 4.00p.m., *O Yang 77* approached Waitangi from the west on a heading of 090 degrees (T). In view of the prevailing weather conditions this meant that the trawler was approaching a lee shore. This, together with the fact that the master was unfamiliar with the area, would normally have meant that the approach would have been made with a fair degree of caution.

The fishing vessel *Lady Iris*, with the fisheries officer and the four joining seaman on board, left the wharf at 5.15p.m., just as the trawler was sighted coming up into the bay. It soon became apparent to those on the fishing vessel that the trawler was approaching rather faster than was considered wise. 'Vessel appeared to be coming in quite boldly,' was how one watcher on board *Lady Iris* described it.

On board *O Yang 77* main-engines were stopped as they approached the waiting *Lady Iris*, then lying approximately half a mile north of the wharf. The mate went forward to the forecastle to let go the starboard anchor, which had previously been walked out to almost water level. Unfortunately, the main-engine was not put to 'full astern' to take the way off the vessel, and on the forecastle the mate found that he was unable to let go the starboard anchor owing to a defective clutch relay switch. With wind and sea pushing her towards the lee shore, the trawler overshot the waiting *Lady Iris*, causing some concern to those onboard. As the fishing vessel tried to catch up with the trawler, they called up on Channel 16, warning the master that he was getting into shallow water and should turn immediately. At this stage it would appear that the main-engine was put to 'full ahead' and the wheel to 'hard-a-port'. The trawler slowly started to turn, but it was too late. As she came round, the westerly wind and swell took her broadside on and set her towards the shore. The watchers on *Lady Iris* saw sand being churned up around the stern and at 5.25p.m. the trawler grounded. The engine was put 'full astern' but had no effect and, after five minutes or so, was shut down. An attempt was then made to pass one of the trawl-wires out through the stern so that it could be run ashore and made fast to the wharf, about 800 metres away, the intention being to kedge the vessel off the beach using the main trawl-winch, or at least prevent her being pushed further inshore. Three local fishing boats tried to run the wire, but it proved too heavy for them to handle and the attempt had to be abandoned. During the night of 15-16 June strong south-westerly winds and a heavy swell drove the stranded trawler 400 metres northwards up the beach and into shallower water.

Southern Tug & Barge personnel and equipment began arriving on Chatham Island on 17 June and, the next day, the tug *Southern Alpha* and our appointed surveyor from Auckland, Chris Laird, arrived. The salvage operation was ready to commence.

For a vessel stranded like *O Yang 77* the following techniques are normally used to achieve refloating:-

(1) Dragging vessel to deep water by means of tension in beach gear anchor systems.
(2) Removal of the ground around and under the hull by dredging, thereby allowing vessel to float inside dredged area. Then dredging a channel to deep water.
(3) Removal of weight, such as cargo, stores and fuel, to allow vessel to float at reduced draughts.

In the case of *O Yang 77*, the trawler was so far out of her floating draughts that it would not be possible for beach gear alone to overcome the ground reaction, friction, and suction forces. Dredging of the ground around the vessel would be required so as to reduce these considerable forces by as much as possible. Further reduction in the force required of the beach gear would be obtained by removing weight in the form of the fish-cargo, stores, and fuel. Therefore, to successfully refloat the trawler a combination of methods (1), (2) and (3) would be required.

The condition of the vessel on 18 June was that the integrity of the hull had not been affected and, apart from sand choking the seawater cooling system, the main and auxiliary machinery appeared to be undamaged, as did the propeller and rudder. The trawler was almost upright but buried in the sand up to her sailing draughts of Forward 3.20 metres, Aft 5.20 metres.

The fouling of the cooling system had caused the two diesel-driven generators to be shut down to prevent overheating. The salvage team therefore placed priority on cleaning the sand from the cooling system and, within 48 hours, both generators were back in service. All the trawler's deck machinery, including the electro-hydraulic cargo winch system, was operational and could, therefore, be used for the discharge of cargo and stores. The main trawl-winch, which could be used in conjunction with beach gear anchors to pull the vessel clear of the ground, was also in working order. This winch was driven by hydraulic power generated by a main-engine-driven hydraulic pump. However, to prevent damage to the propeller while using the trawl-winch it would be necessary to disconnect the main-engine from the propeller shaft.

On 19 June the salvage crew commenced rigging up sheerlegs and a jack stay for a 'flying fox' system to facilitate the discharge of cargo and stores to the beach. At the same time, the vessel's anchors were laid out to seaward and the cables heaved tight to prevent further movement up the beach.

Discharge of the fish-cargo (estimated at approximately 100 tons), the dry stores and some 280 tons of fuel oil commenced on 20 June. At the same time, a general inspection of both the casualty and the surrounding area indicated that additional divers would be required, together with anchors and wires, so that at least four sets of beach gear could be laid out to seaward of the casualty. With the trawler buried in the sand up to her sailing draughts, the other essential equipment that would have to be transported to the island was sand dredging and extraction machinery. It was also realised at this stage that the services of an experienced salvage officer would be required to co-ordinate the salvage operation effectively. To this end Southern Tug & Barge engaged the services of Captain David Hancox, an old friend of ours from many past salvage jobs.

Mobilisation of the additional men and equipment started on 23 June and, on 28 June, the tug *Parahaki* arrived at Waitangi wharf from Wellington and commenced the discharge of the additional beach gear, sand dredging and extraction machinery, and diving equipment. Next day Captain Hancox arrived and took overall charge of the operation.

The salvage plan was basically to remove as much weight as possible from the vessel in the form of fish-cargo, stores and fuel, to lay out at least four sets of beach gear to seaward of the casualty and heave the anchor wires bar tight. At the same time, the sand around the casualty would be dredged away. The expected weather conditions were strong south-westerly winds and a heavy swell – the very conditions that put the trawler on the beach. It was anticipated that the swell, together with the dredging of the surrounding area, would provide the vessel with sufficient buoyancy for the tension in the beach gear legs to pull her slowly away from the beach towards the deep water. Unfortunately, the period following the stranding was characterised by a spell of unusually calm weather. This, while assisting the removal of cargo and stores, and the work of the divers, did not provide the buoyancy that was required to enable the beach gear to overcome the ground reaction, static friction and suction, thereby dragging the trawler towards deep water. One of the salvage team colourfully, and rather optimistically, described the situation in these terms: 'We've just got to get the environmental conditions stacked up in our favour and away she'll go.' Unfortunately, environmental conditions very often seem to have an inherent reluctance to stack themselves in the favour of those most in need.

Another problem encountered was the structure of the Waitangi beach. This was found to consist of fine sand with extensive layers of peat at various depths beneath the sand. The removal of the sand from around the hull, using the dredging and extraction equipment, was no great problem. However, the removal of peat proved to be not nearly so easy.

From 29 June to 9 July a total of four sets of beach gear legs were laid out to seaward of the stranded trawler. Each beach gear leg consisted of an anchor connected to 100 fathoms of 40mm wire rope, which was led back to the deck of the casualty and connected to the running block of a four-sheave purchase system, the standing block of which was secured to the deck structure. A 16mm wire rope was reeved through the blocks and onto the drum of one of the deck winches. With the winch heaving on the purchase block wire it was estimated that a pulling force of approximately 35 tons would be exerted on the beach gear anchor. With four sets of beach gear deployed we would, therefore, have a total pull of approximately 140 tons.

Once the beach gear was in position and tensioned to hold the vessel, the trawler's port and starboard anchors were recovered and brought back on board. By this time the discharge of the fish-cargo, stores, and fuel had been completed. The frozen fish had been placed in various local freezers around the island, as there was no single freezer that could accommodate the entire cargo. Of the 360 tons of fuel on board at the time of the stranding, 250 tons had been discharged into storage tanks ashore, and 30 tons transferred to the tug *Parahaki*. Approximately 80 tons were retained on board for use in the main and auxiliary machinery during the course of the salvage.

Following his initial assessment of the situation, Captain Hancox decided that, in order to increase the pulling power of the beach gear, a flat-top barge fitted with heavy-duty deck winches be mobilised from Wellington. In view of the anticipated problems with dredging the peat from around the casualty, heavy earth-moving equipment was loaded on the barge prior to departure.

The barge *Pacific* arrived alongside the Waitangi wharf on 16 July and immediately began discharging the earth-moving equipment and additional beach gear. The barge was then towed to a position seaward of the casualty and connected to both the casualty and the four sets of beach gear. During the period 21-28 July attempts to refloat were made by increasing the tension in the beach gear legs on each high tide. Initially, the vessel moved about 20 metres forward; after that, nothing. On most days the weather remained fairly calm, with no swell coming into the bay to help provide that extra buoyancy needed to overcome the ground reaction and the other friction forces.

Inspection of the area round the stranded trawler showed that she was lying on a bank of peat with the box-shaped keel firmly embedded in the peat over the vessel's full length. It was obvious to everyone concerned that, unless the keel was freed from the peat, the beach gear anchors would not be able to move the vessel. In an attempt to overcome this problem a local slipway trolley was modified so that a mechanical digger could be mounted on it. The trolley was fitted with large-diameter aircraft tyres, which enabled it to work in up to eight feet of water and dig away at the peat at most stages of the tide. Because of the difficulty in getting at the peat immediately under the vessel, it was decided to excavate an area astern of the casualty, the intention being to pull the casualty into the dredged basin stern first. While the keel was stuck in the peat and sideways movement was proving impossible it was hoped that, by heaving astern, using the trawl winch connected to two deadmen on the shore, the keel would slide out.

Arrangements were completed by 4 August and on the evening high tide the vessel was moved astern by about six metres. However, on the following morning the weather suddenly deteriorated and strong winds and a heavy swell forced the casualty across the dredged basin to lie hard aground on the landward side. The situation had not improved.

The date was now 5 August and the salvage operation had been going on for 48 days. The daily hire charges were increasing, and we had very little to show for it in the way of progress. Our initial fears about the dangers of a long, drawn-out salvage operation on a daily-rate contract were being realised. Our costs were now more than NZ$2m., and in Singapore we were getting more and more 'flak' from the underwriters in Seoul, who could not understand why it was proving so difficult to refloat the trawler. So 7 August found me in Waitangi looking at the situation first-hand and discussing the various options with Captain Hancox and our local man, Neil Abbot   Chris Laird having returned to Auckland owing to other commitments.

The position was that after 48 days' work and ever-increasing costs – two tugs, a salvage barge, bulldozers, diggers and diving teams do not come cheap – we were no nearer refloating the casualty than we had been on 18 June. Initially, it had been hoped that the casualty would be refloated quite quickly. However, it had now become appar-

232

ent, in view of the many problems encountered – not least being the logistics associated with the remote geographical location of Chatham Island – that to continue the operation under the original contract terms would only result in an ever-increasing salvage bill. After much discussion it was agreed to adopt the following procedure:-

(1) Suspend current operations to refloat casualty and demobilise all shore-based and floating equipment, except that which was immediately necessary to maintain the trawler in a safe condition.
(2) Invite local and overseas salvage contractors to submit Lump Sum No Cure – No Pay tenders to refloat vessel and redeliver to the owners at Lyttelton.
(3) Maintain casualty in present position under the charge of the salvage officer and five-man maintenance crew to tend the beach gear and refloat the vessel, should a favourable situation arise.

Demobilisation of the heavy equipment commenced on 10 August and the two tugs and the flat-top barge, loaded with the excavator, bulldozer and diving equipment, departed Waitangi for Wellington on 14 August.

The weather remained calm until 15 August when a south-westerly wind and swell started to pick up. This was exactly the weather we had been hoping for since the salvage operation started. Tensioning of the beach gear during the evening high tide, combined with the buoyancy provided by the swell, resulted in a movement of 30 metres towards deeper water. Similar weather continued through the next day and a further movement of 15 metres was achieved. Unfortunately, at this stage a previously-arranged business commitment forced Captain Hancox to return to Australia. However, Southern Tug & Barge had managed to obtain the services of Captain Charles Deeney, another

233

very experienced salvage officer, to take over the reins from David Hancox. Charlie Deeney was also an old friend of ours whom I had first met on a salvage job at Saint John, New Brunswick, in Canada, twelve years previously.

During my time on Chatham Island, I stayed in the Waitangi Hotel. In the evenings the hotel bar was always fairly busy with 20 or 30 of the local fishermen and farmers drinking the local ale, Black Robin, which is not a bad brew at all, and playing darts. The stranded Korean trawler was the subject of much conversation and many were the theories put forward for the successful refloating of the distressed vessel. Some were quite fanciful; some not so fanciful – not that we could afford to be very critical since, after about 50 days and the expenditure of $2m., we had absolutely nothing to show in the way of success. So in the bar we kept our profile low and our ears open.

The local ale is named after a native species of bird found nowhere else. As the name suggests the black robin is very similar to the European robin, except that it is totally black. For one reason and another the black robin of 'The Chathams' became an endangered species and in 1976 the number of birds had dropped to about seven. Fortunately, an excellent operation by the New Zealand Wildlife Service saved the robins from extinction and, by 1990 the numbers had increased to more than 100. Mind you, their numbers would be in the thousands if their survival was linked to the amount of Black Robin ale consumed nightly in the bar of the Waitangi Hotel.

However, to get back to the business of our stranded trawler, the weather was calm during 17-18 August, but on 19 August the south-westerly wind returned with rough seas and a heavy swell. Some further movement towards deep water was achieved, but heavy pounding in the rough seas caused the operation to be suspended, the casualty having to be ballasted down and secured in order to prevent damage to the bottom. Next day, the weather abated some-what but there was enough of a south-westerly swell to provide the buoyancy required to help move the casualty away from the beach; tensioning of the beach gear at the top of the tides resulted in further movement, metre by metre, towards deep water.

21 August saw the south-westerly swell continue to run up into the bay and, as the swell provided the buoyancy necessary to reduce the ground reaction, skilful control of the beach gear by Charlie Deeney and the salvage crew enabled the casualty to continue her movement away from the beach. At first, the rate of progress was quite slow – less than a metre each time a swell provided lift to the casualty. Then the rate of movement steadily increased to two or three metres with each passing swell. As the deck winches exerted maximum load on the beach gear, one of the beach gear anchor wires parted under the strain, but the others held. Then, suddenly, at 4.00p.m. *O Yang 77* came free of the ground and moved into deep water.

Work immediately commenced, reconnecting the main-engine to the propeller shaft and, at daylight on 22 August, an underwater inspection of the hull was carried out. This revealed that, despite the 67 day stranding on the beach, no obvious hull damage had been sustained, apart from the crushing of the echo-sounder transducer boxes. At this stage the vessel was lying safely secured by the three beach gear anchors. During the day the wind increased, with a south-westerly wind of 25 knots gusting to 40 knots, accompanied by a heavy swell. This caused another of the beach gear anchor wires to part, but the other two held.

On the morning of 23 August the beach gear anchors were slipped and the trawler proceeded without assistance to the Waitangi wharf where the stores, equipment and fuel off-loaded during the salvage operation were reloaded. At the same time, a check of all essential systems was carried out to ensure the vessel was in a fit condition to undertake the passage to Lyttelton under her own power.

The voyage preparations were completed on 25 August and at 2.15p.m. the trawler departed Waitangi. The passage was uneventful and *O Yang 77* arrived safely alongside the berth at Lyttelton at 5.30p.m. on 27 August. Demobilisation of the salvage crew and equipment commenced that evening and was completed at 11.30a.m. on 28 August, at which time the trawler was redelivered to its owners, 75 days after running aground at Waitangi.

The final salvage costs amounted to NZ$2,928,550 – almost NZ$40,000, per day. However, the good news was that inspection on drydock at Lyttelton revealed that the underwater hull was, apart from some minor scores and indents, undamaged.

After spending hundreds of thousands of dollars on barges, heavy digging equipment, bulldozers, tugs, diving teams and dredging gear, refloating had, in the end, been achieved simply by the use of the beach gear anchors and the heavy south-westerly swell. As some wag had remarked about seven weeks earlier, 'Just a matter of waiting until environmental conditions finally stack up in our favour.' If only we had had that swell at the start of the operation . . . but I am afraid that's the salvage business for you.

One would have thought that, after all the bother that they got themselves into at Chatham Island, the owners and crew would have been very careful with their vessel thereafter. Alas, this was not so.

While lying alongside the berth at Lyttelton, New Zealand, on 28 January 1995, taking on diesel fuel prior to departing for the Pacific fishing grounds, a fire occurred, at 6.00p.m., in one of the six-berth crew cabins on the lower accommodation deck. The fire, which, apparently, had originated at a wall-mounted fluorescent reading-light, spread to the plywood bulkhead linings in both the crew cabin and adjacent spaces. The local fire brigade was called, and the fire brought under control and finally extinguished next day.

The vessel's underwriters on this occasion were Samsung Fire & Marine Insurance Co., Seoul. Our old friends, Ankuk Fire & Marine, had perhaps recognised that this vessel represented an unacceptably high risk. If so, they were not far wrong. Anyway the new underwriters instructed us to attend on their behalf and report on the cause, nature, and extent of the damage, and the estimated cost of repairs.

Our survey showed that the damage affected the wheelhouse, the upper and lower accommodation spaces and the domestic reefer spaces. The internal fittings in these spaces had all suffered either fire or water damage, and the hull plating on the port and starboard sides of the damaged area was distorted by the effects of the fire.

The vessel was insured on a 'Total Loss' basis only for US$3m. Our estimate of the repair cost, including a substantial amount for possible unknown additional items that might be revealed during the course of repairs, came to NZ$3,566,250. (US$2,310,880.) This was below the insured value, which meant that the owners would have to pay the cost of repairs. They were not happy with this situation and two months of somewhat heated discussions followed. We were, however, quite satisfied that our estimate accurately represented the fair and reasonable cost of the fire damage repairs and the underwriters were advised to that effect.

Eventually, towards the end of April, the Owners carried out temporary repairs at Lyttelton and sailed the vessel back to Korea where they would be able to carry out permanent repairs for a much lower cost than in New Zealand.

We were never advised of what happened after the trawler returned to Korea, perhaps she was repaired, perhaps not. But, hopefully *O Yang 77* still hunts the Pacific fishing grounds, supplying the fish markets of the world with yellow-fin tuna and other fruits of the ocean.

13

The Road to Mandalay

Come you back to Mandalay,
Where the old Flotilla lay:
Can't you 'ear their paddles chunkin'
   from Rangoon to Mandalay?
On the road to Mandalay,
Where the flyin'-fishes play,
An' the dawn comes up like thunder
   outer China 'crost the Bay!

In the final year of my engineering apprenticeship with John G. Kincaid & Co., Ltd at Greenock I worked on the assembly of the diesel engine fuel injection pumps. This work was carried out by an elite little group of two fitters and two final year apprentices. The senior fitter was Willie Taylor, a really first-rate man. To me he was the archetypal Clydeside engineer; a master craftsman who made even the most tricky task look ridiculously simple. In later years, when confronted with some awkward or complicated piece of machinery to dismantle or assemble, I would invariably think to myself, 'now how would Willie have gone about this?' Strange as it may seem, it was Willie who really introduced me to the poems of Rudyard Kipling for I don't remember reading any of Kipling's work at school. One day, after yarning about his sea-going days in the Far East and telling me about his favourite poem, *McAndrew's*

Hymn, which must surely have been inspired by some Scottish chief engineer who Kipling met on one of his ocean travels, Willie gave me a little red book of Kipling's poems entitled *The Seven Seas*. That little book has been all over the world with me; nearly 50 years on it sits in my bookcase in Perthshire. Willie, I fear, has long since departed this world of ships and engines, but I warm to his memory still.

I think it was Kipling more than anything else that sparked off in me a lifelong fascination with the East. His poem *Mandalay* was my favourite, the very name Mandalay appeared to conjure up all the magic of the Orient and, in my mind's eye, I always had a vivid picture of 'the dawn coming up like thunder'. However, I was puzzled by the words 'the old Flotilla'. I realised that this referred to the river steamers and barges of the Irrawaddy Flotilla Company that operated up and down the river between Rangoon and Mandalay, but it seemed a strange name for a commercial company. Flotilla to me indicated a naval or military influence. Several years later, while working for P. Henderson & Company, the Glasgow ship owner who had a close relationship with both Burma and the Flotilla Company, I learned something of the linked history of these famous old Scottish companies and the origins of the name Flotilla Company.

The Irrawaddy Flotilla Company had its origins in the second Anglo-Burmese War of 1852 when the Governor General of India ordered the Indian Marine to send a flotilla of four paddle-steamers and four flats from Calcutta to Rangoon. The purpose of the flotilla was to transport troops and equipment, and maintain a line of communication on the Irrawaddy river. The paddle-steamers and flats provided valuable service both during the campaign and subsequently, when they were used to maintain garrisons stationed up and down the river. The flats, which accompanied the steamers, were the hulls of older steamers from which engines, boilers and upper-works had been removed, leaving clear decks to accommodate troops and military equipment.

Following the war of 1852 the importance of the Irrawaddy as the main communication artery of Burma became increasingly obvious and it was decided that the flotilla should be expanded. However, in order to take advantage of mercantile trade on the river as well as transporting government personnel and equipment, it was thought best that the operation of the flotilla should be carried out by a commercial organisation.

The Government of India offered the flotilla for sale in 1865 and it was purchased by Todd Findlay & Co. of Glasgow for £16,200. Todd Findlay was engaged in the export of rice and teak at Moulmein and Rangoon and, in addition, operated four paddle-steamers on a coastal service between Moulmein and Victoria Point on the Tenasserim coast. They had a close relationship with the Glasgow shipping company, P. Henderson & Co, who, as described in an earlier chapter, had been trading their sailing ships down to New Zealand from about 1854. At that time, return cargoes from New Zealand were scarce but the Henderson ships found a profitable return business by running in ballast up to Burma, where Todd Findlay supplied them with homeward cargoes of rice and teak at Moulmein and Rangoon. The success of this trade led the Henderson Line to open a regular direct service between Glasgow and the two Burmese ports in 1860, providing a permanent link between them and Todd Findlay.

Following the purchase of the flotilla, Todd Findlay found that the operation and expansion of the fleet required more capital than they could provide. They therefore turned to their friends at the Henderson Line for assistance; as a result a syndicate, consisting of Todd Findlay, William Denny (the Dumbarton shipbuilders) and the Henderson Line, was formed to take over the running of the river fleet.

The registered office of the new Irrawaddy Flotilla Company was at Henderson's office in Glasgow, with James Galbraith, the Henderson Line senior partner, as managing director.

Rudyard Kipling made his acquaintance with the Flotilla Company during the third Anglo-Burmese war of 1858, when almost the entire fleet was requisitioned by the government to transport the Burma Field Force from Rangoon up the Irrawaddy to Mandalay.

The river trade outgrew all expectations and, with additional senior staff sent out from Henderson's Glasgow office, the fleet continually expanded. New river steamers were built mainly at Denny's yard at Dumbarton, part-dismantled and shipped to Rangoon on Henderson Line ships to be re-assembled at the Dalla shipyard on the opposite side of the river from Rangoon. By the end of the century the fleet consisted of 200 steamers and over 300 flats and cargo barges, operating from Rangoon and Bassein on the delta up past Mandalay to Bhamo, close to the border with China – a distance of more than 1,000 miles. The Irrawaddy Flotilla Company had become the largest inland water transport operation in the world.

During the Second World War most of the Flotilla Company's fleet, then 650 vessels, were either destroyed by enemy action or scuttled to deny them to the invading Japanese forces. Then, after General Slim's 14th Army had driven the Japanese out of Burma in May 1945, work commenced on restoring the most essential of the river services and, by the end of the year, the old Flotilla Company was back in business. It was, however, fated to have a limited future. In June 1948 the newly independent Burmese Government nationalised the Flotilla Company and re-named it The Inland Water Transport Board. It was the end of an era.

In an earlier chapter I have described my first visit to

Rangoon more than 45 years ago, when I was 2nd engineer on the Henderson Line cargo ship *Bhamo*. Rangoon did not disappoint me as I gazed in awe at the massive and ornate Shwedagon Pagoda. Lying at our river moorings off Sule Pagoda wharf, loading our homeward cargo, I watched the paddle-steamers and barges pass close by, coming from and going to Mandalay and the other Irrawaddy river ports. What an adventure it would be, I thought, to sail on one of them all the way past Bagan and Mandalay right up to Bhamo.

Over the next 40 or so years I made frequent visits to Burma, firstly as a Henderson Line engineer, then as a marine surveyor with Ritchie & Bisset and, finally, with the Salvage Association. During the last few years of my service with the Salvage Association I was often amused when Maung Maung Lay, our representative at Rangoon, invariably introduced me to senior Burmese Marine Department officers with the words: 'Mr Walker is an old Henderson Line man, you know'. I rather liked it, and suitable reverence was always shown! Many of the Marine Department officers had, in fact, spent their early days as apprentices and junior officers on Paddy Henderson's ships and so we were able to reminisce about captains and chief engineers that we had known. I am sure they enjoyed our nostalgic discussions every bit as much as I did for they hadn't seen a Henderson ship in Rangoon since 1967, so for them to meet one of Paddy's old hands in the late '90s was a bit of a novelty.

However, despite trying to find an excuse to visit Mandalay and watch Kipling's dawn, I was never able to penetrate beyond the port cities of Rangoon, Bassein and Akyab. Then, in January of 1997, just before I retired from the Salvage Association, my wife and I were able to travel on the new Orient Express river cruiser *Road to Mandalay* from the ancient temple city of Bagan to Mandalay.

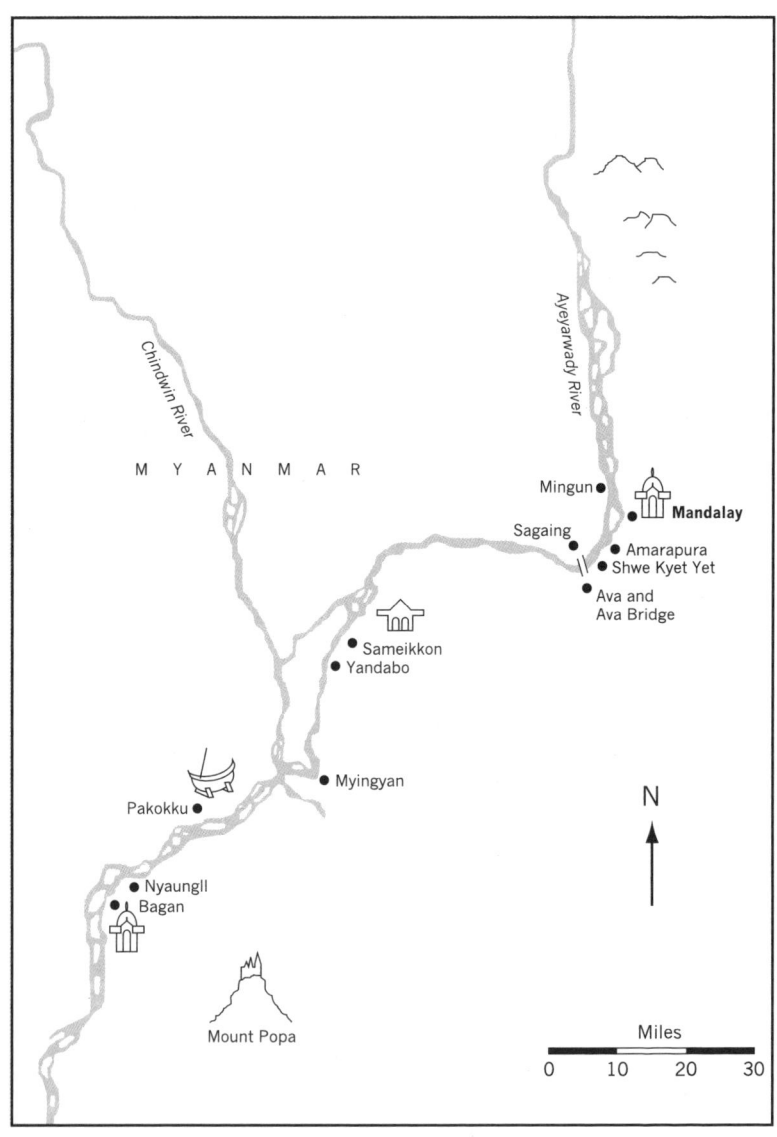

Map 18 Irrawaddy (Ayeyarwady) River Bagan to Mandalay

To arrive in Bagan is to travel back in time hundreds of years. There is an air of complete and utter serenity, exemplified for me in the slow, plodding, ox-driven carts. Approximately 13,000 pagodas and temples were built at Bagan in the 11th century. Many have been destroyed by erosion as a result of the Irrawaddy or Ayeyarwady – as it is now officially known – changing its course over the centuries, the effects of earthquakes and the ravages of time. However, approximately 2,000 remain and, as far as the eye can see, there is nothing but pagodas and temples. Some are complete and functioning as places of worship, some have been damaged by the passage of time, and some are just heaps of bricks. Our dainty Karen guide whisked us round various temples on the afternoon of our arrival, finishing by taking us to watch the sunset from high on the ramparts of the That-Byin-Nyo Temple. The view was spectacular: ox-carts trundling homewards, temple bells ringing, smoke from cooking fires rising gently in the dusty air and the pink, golden and crimson rays from the setting sun gilding the spires and towers.

Next morning, after an early breakfast, we were shown round the Shwe-zigon Pagoda, stated to be one of the most famous of all Bagan's monuments. We then visited the market at Nyaung U, some eight kilometres from Bagan. The market was fascinating with an amazing variety of fresh fruit and vegetables from the Shan hills on sale, together with embroidered work, puppets and lacquer-ware. But the most interesting aspect of the market was presented by the people of the hill tribes who had come down to sell their wares – Shans, Chins and Kachins, all in their distinctive, colourful, tribal garb.

We were back on board by 11.00a.m., and, shortly afterwards *Road to Mandalay* un-berthed and sailed. As we moved out into the river, what appeared to be a line of fishing stakes could be seen sticking out of the water at an angle of about 45 degrees, each with a tin reflector attached at the upper end. Closer inspection, however, revealed these to be the buoys marking the navigable channel. They consist of long bamboo poles from five to eight metres in length, depending on current strength and depth of water, the poles being anchored to the river bottom with sand-filled gunny-bags. The tin discs at the upper ends are to assist navigation at night by reflecting the steamers' searchlight beams. In January the river level is fairly low, although it is usually at its lowest level in March, and survey launches are employed every day at critical stretches of the river, checking water depths with bamboo sounding-poles and then adjusting the position of the marker buoys as found necessary. When passing the survey launches, and also some of the local river craft, there was much shouted discussion between them and our pilot regarding the exact depth of water that day. I imagine that is exactly how it was back in the days of the old Flotilla Company.

For the rest of the day we sailed up-river, slipping silently past whitewashed temples, small villages, patient oxen – life by the river bank unchanged for centuries. On the river itself, large rafts of teak logs and bamboo passed slowly down-river towards Rangoon, each raft having little bamboo huts for the 'crew'. On the sand-banks exposed by the falling river level of the dry season were numerous temporary villages, housing an itinerant population engaged in loading sand into country river-craft which is then sold to make cement. Just before sunset, we anchored in midstream near Myingyan, close to where the Chindwin river joins the Irrawaddy. The sunset was absolutely magnificent – a kaleidoscope of colours: yellow, gold, crimson, pink, grey, blue and black.

Next morning we weighed anchor just after sunrise and continued up-river towards Mandalay, where we arrived in the early afternoon. The approach to Mandalay was quite spectacular as our ship slid under the Ava bridge – the only bridge to cross the Irrawaddy – with the ridge of the Sagaing Hills on the port side dotted with white- and gold-spired pagodas and, on the starboard side, the picturesque Shwe Kyet Yet pagodas.

However, for me by far the most impressive part of the river passage from Bagan had been the early morning, as we prepared to weigh anchor for Mandalay – a pale, silvery morning with a few shivering passengers on deck looking eastwards to a horizon already tinged with gold and orange. Suddenly, a sliver of bright red appeared and, with perceptible movement, the sun climbed towards the cloudless sky. With a feeling of awe and wonderment I achieved my lifelong ambition of watching the dawn come up 'like thunder . . . On the road to Mandalay.'